T0262305

Modelling and Control of Air Pollution

Modelling and Control of Air Pollution

Edited by **Raven Brennan**

New York

Published by Callisto Reference,
106 Park Avenue, Suite 200,
New York, NY 10016, USA
www.callistoreference.com

Modelling and Control of Air Pollution
Edited by Raven Brennan

© 2015 Callisto Reference

International Standard Book Number: 978-1-63239-461-3 (Hardback)

Printed in the United States of America.

Contents

Preface

The crucial problem of air pollution possesses a universal nature. It has been a major environmental problem and an issue of global interest for many years. High concentrations of air pollutants due to several anthropogenic actions influence the quality of the air. This book discusses various topics like the impact of air pollutants on air quality, the debilitating effects of industrial emissions, among others. It presents a variety of monitoring techniques of air pollutants, their predictions and control. It also involves case studies explaining the exposure and health implications of air pollutants on living beings in various countries around the world. The book will be of great help to graduates, professionals and researchers.

After months of intensive research and writing, this book is the end result of all who devoted their time and efforts in the initiation and progress of this book. It will surely be a source of reference in enhancing the required knowledge of the new developments in the area. During the course of developing this book, certain measures such as accuracy, authenticity and research focused analytical studies were given preference in order to produce a comprehensive book in the area of study.

This book would not have been possible without the efforts of the authors and the publisher. I extend my sincere thanks to them. Secondly, I express my gratitude to my family and well-wishers. And most importantly, I thank my students for constantly expressing their willingness and curiosity in enhancing their knowledge in the field, which encourages me to take up further research projects for the advancement of the area.

Editor

Pollution and Air Quality in Târgovişte Municipality and Its Surroundings (Romania)

Pehoiu Gica and Murărescu Ovidiu
"Valahia" University of Târgovişte
Romania

1. Introduction

Târgovişte Municipality is situated in the High Plain of Târgovişte, at an average absolute altitude of 280 m. The town has an administrative area of 4,681 ha, in which the constructible area includes 1,966 ha; within the latter, 100.7 ha represent green area (Fig.1).

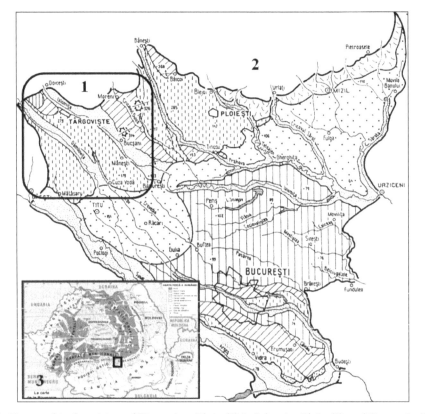

Fig. 1. Geographical position of Târgovişte Plain (1) in Ialomiţa Plain (2) and Romania (3).

Târgovişte had a population of 88,119 inhabitants in the year 2010 - representing 54.06% of the total urban population of Dâmboviţa County and 16.63% of the total population of the county -, the density of the town's population being of 1,882.4 inhabitants/km² (Statistical Yearbook of Dâmboviţa County, 2010).

This old industrial center has developed and diversified concomitantly to the general development of the economy (after the year 1968, on the occasion of the new administrative-territorial division, Târgovişte became political-administrative center of the county). It is during this period that the industrial platform, with several economic units and new branches appeared, completing the town's industrial profile; this profile has undergone significant changes after the year 1990. Here, at present, there are nationally important industrial units, on the platform situated in the south-west of the town. They actually represent the main sources with a potential impact on the air quality in Târgovişte municipality and its neighborhood. Among these, one can mention: SC Mechel SA, in the metallurgical domain, producing special steels, SC Upet SA (now being restructured), SC Nemo SA, focused on machine building, SC Swarco - Vicas SA, producing paints and varnish, SC Oţel Inox SA, laminating steels, SC Cromsteel SA, a company based on chroming processes, SC Romlux SA, an industrial unit producing light fixtures and others (Pehoiu, 2003; Pehoiu et al., 2005).

Doiceşti commune is situated near Târgovişte municipality; this commune also holds some industrial units, of which Uzina Electrică (the Power Station), using black oil and brown coal as fuels to generate electricity, SC Nubiola România SRL, a company producing whiteners and green chrome oxide - SO_2 and possibly a sulfuretted hydrogen source in the process of sulphur combustion needed to obtain ultramarine -, SC Soceram SA, a company producing bricks and ceramic materials.

Fieni Town, situated in the north of the political-administrative center of Dâmboviţa County, is remarkable through its industrial units: SC Carpatcement Holding SA, a producer of building materials (cement) and SC Carmeuse Holding SRL (lime producer), which represent the main sources of environmental degradation for the locality of Fieni and its surroundings (Pehoiu, 2008).

Târgovişte Town is a significant consumer of resources and at the same time a major producer of polluting emissions, resulting mainly from: industrial activity, intensification of road traffic and generation of high quantities of waste.

2. Working methodology

In order to analyze the air quality status and the effects of the air pollution in a mainly industrial town that has gone through important structural modifications after 1990, when we analyzed the pollutants' dispersion, we took into account as well the role of the climatic factors (wind – its speed and directions, atmospheric humidity, atmospheric calm, hydrometeors - fog, air temperature).

The air analysis network component pertaining to Dâmboviţa County is managed by the Departmental Agency for Environmental Protection - AEP (Agenţia Judeţeană pentru Protecţia Mediului) and includes fixed monitoring stations in the localities Târgovişte (Fig. 2) and Fieni (an automatic station each), and one in Doiceşti, with manual sampling of the pollutants and analysis in the laboratory, by means of which the specialists monitor the concentrations of the dusts in suspension and of the gaseous pollutants (ammonia, nitrogen oxides, sulphur dioxide, sulphuretted hydrogen, formaldehyde).

Fig. 2. Map of the air monitoring locations in Târgovişte municipality (AEP Târgovişte).

At present, the monitoring of the air quality in these points supposes the continual gathering of daily samples from the atmosphere (24 h), followed by the analysis of the samples in the laboratory. This kind of analysis allows to highlight the dangerous concentrations for the population's health in due time. The data obtained following the measurements serve to create databases and to elaborate reports or informative bulletins in the aftermath of the occurrence of eventual pollution episodes.

Excepting the meteorological parameters, a series of polluting indicators were monitored, such as: benzene, carbon monoxide, sulphur dioxide, nitrogen oxides, lead, heavy metals etc. These pollutants were gathered from several manual sampling points distributed around the town, including the residential and the industrial areas.

In order to determine the quantities of dusts and polluting emissions present in the atmosphere and influencing the air quality status, the polluting agents were separated depending on the area they influence (for instance, the dusts emitted in the atmosphere by a series of metallurgical companies are carried over two residential quarters in the south-east of the town.

Following the comparative analysis of the polluting components, during the last few years one can notice a slight improvement of the air quality compared to the previous period in point of dusts-caused pollution. In the areas in which they are monitored in the atmosphere, gaseous pollutants (nitrogen dioxide, sulphur dioxide, ammonia, sulphuretted hydrogen, formaldehyde, oxidant substances, and carbon monoxide) are not present in concentrations over the limits allowed by the present legislation.

The pollutants to be monitored, the measurement methods, the limit values (LV), the alert and information thresholds and the criteria for situating the monitoring points are established by the national legislation concerning the atmospheric protection, being in agreement with the demands of the European regulations.

The realizations concerning the development of the air quality monitoring network in Dâmboviţa County during the period 2007-2010 consisted in:

- remodeling and adapting the environmental laboratories for the installation of new laboratory equipments; installing these equipments; instructing the personnel;
- completing the endowment of the automated air quality monitoring stations from Târgovişte and Fieni.

AEP Dâmbovița monitored the level of the dusts in suspension in the area of Târgovişte, by means of its four sampling stations. In the municipality, the indicator of breathable dusts was monitored in the PM_{10} fraction in the quarter Micro XII (2 representative points for the dusts resulted from SC Mechel SA Târgovişte) and the Civic Center – representative for the dusts resulted from road traffic and other sources after dispersion (SC Mechel SA) – where the indicator monitored was that of total dusts in suspension.

3. Air quality

The air is the environmental factor constituting the most rapid support favoring the pollutants' transportation in the environment. Air pollution has many and significant negative effects on the population's health and may damage as well the flora and fauna in general.

Air quality depends on the emissions that end up in the air coming from stationary and mobile sources (road traffic), mainly in big cities, as well as on the pollutants' transport on longer distances.

The automated stations and the manual sampling points are situated in representative areas in point of pollution, in the localities mentioned, as follows:

- *Automated station DB-1*, situated in Târgovişte municipality, Vlad Ţepeş Str., no. 6 C (in the courtyard of the Social Care Center "Sfânta Maria"), coordinates: 25⁰28′41.6′′; 44⁰54′58.39′′;
- *Automated station DB-2*, situated in Fieni Town, Teilor Str., no. 20 (in the town's central park), coordinates: 25⁰25′18.30′′; 45⁰07′52.98′′;
- *Fixed manual sampling point* in Târgovişte - PM_{10} fraction, in the Micro XII station, point 1, Constructorilor Str., no. 21 (Procor headquarters, industrial platform);
- *Fixed manual sampling point* in Doiceşti (CFR Station) – gaseous pollutants (sulphur dioxide - SO_2, nitrogen dioxide - NO_2, sulphuretted hydrogen - H_2S and hexavalent chromium – expressed as CrO_3) and total dusts in suspension (TSP). The gaseous pollutants determinations in the point Doiceşti continue to be carried out in agreement with the analysis methods mentioned by STAS 12574/1987.

Settleable dusts are monitored by means of measurements in six fixed points situated in the following locations:

- Târgovişte - 3 sampling points – point 1 Micro XII, AEP headquarters, and Micro XI (DB-1 station);
- Doiceşti - 1 sampling point - CFR Train Station;
- Fieni - 2 sampling points - point 1 (Fieni Park) and point 2, DB-2 station.

3.1 Atmospheric pollutants

- **Benzene**

General features:

-   - Very light aromatic compound, volatile and water-soluble;
-   - 90% of the benzene present in the air comes from road traffic;
-   - The rest of 10% comes from fuel evaporation during its storage and distribution.

Effects on heath: carcinogenic substance, classified into the A1 toxicity class, known as carcinogenic for man. It produces negative effects on the central nervous system.

Measurement methods: the method of reference for measuring benzene is that of sampling by aspiration into an absorbing cartridge, followed by gas-chromatographic determination, standardized at present by the European Committee for Standardization (CEN).

- **Carbon monoxide**

General features. At room temperature, carbon monoxide is a colorless, odorless, tasteless gas, of both natural and anthropic origin. Carbon monoxide is formed mainly through the incomplete burning of fossil fuels.

Natural sources: forest burning, volcanic emissions, electric discharges.

Anthropic sources: it is formed mainly through the incomplete burning of fossil fuels, steel and pig iron production, oil refinement, and road, air and railroad traffic.

Carbon monoxide can accumulate up to a dangerous level especially during the period of atmospheric calm during winter and spring (this gas being much more stable from a chemical viewpoint at low temperatures), when the burning of fossil fuels attains a maximum level. Produced by natural sources, it is very rapidly dispersed in a wide area, and consequently does not affect human health.

Effects on people's health. It is a toxic gas, being lethal in high concentrations (at concentrations of about 100 mg/m^3) through the reduction of the blood's capacity to transport oxygen, with consequences on the respiratory and cardiovascular system.

At relatively low concentrations, it affects the central nervous system, weakens the pulse rate, diminishing the blood volume distributed in the organism, and at the same time it reduces visual acuity and physical capacity. Being exposed for a short period of time, one may experience acute fatigue. At the same time, it may trigger respiratory difficulties, chest pains in people with cardiovascular diseases, determining as well irritability, migraines, rapid respiration, lack of coordination, nausea, dizziness, confusion, and can reduce the ability to concentrate.

The population segments most affected by the exposure to carbon monoxide are represented by children, elderly, people with respiratory and cardiovascular diseases, anemic people and smokers.

Effects on plants. At concentrations normally encountered when monitoring the atmosphere, carbon monoxide does not affect the plants, the animals or the environment.

Measurement methods: the reference method for measuring carbon monoxide is the non-dispersive infrared (NDIR) spectrometric method: ISO 4224.

- **Sulphur dioxide**

General features. Sulphur dioxide is a colorless, bitter, non-flammable gas, with a penetrating odor that irritates the eyes and the respiratory system.

Natural sources: volcanic eruptions, marine phytoplankton, bacterial fermentation in the marshy areas, oxidation of the gas containing sulphur resulted from biomass decomposition.

Anthropic sources: population's heating systems, when the fuel used is not methane, thermoelectric power stations, industrial processes (siderurgy, refinery, sulfuric acid production), cellulose industry and, to a lesser extent, the emissions coming from diesel engines.

Effects on people's health. Depending on its concentration and the period of exposure, sulphur dioxide has different effects on human health. The exposure to a high concentration of sulphur dioxide during a short period of time can cause severe respiratory difficulties. Particularly affected are: people with asthma, children, elderly and people with chronic respiratory diseases. The exposure to a low concentration of sulphur dioxide for a long lapse of time can result in infections of the respiratory system. Sulphur dioxide can interfere with the dangerous effects of the ozone.

Effects on plants. Sulphur dioxide clearly affects many plant species, the negative effect on their structure and tissues being visible with the naked eye. Some of the most sensitive plants are: pine, vegetables, red and black acorns, white ash, lucerne, blackberries.

Effects on the environment. In the atmosphere, it contributes to the acidification of the precipitations, with toxic effects on the vegetation and on the soil. The increase of the sulphur dioxide concentration accelerates metals' corrosion, because of the formation of acids. Sulphur oxides can erode: stones, brick-and-mortar, paints, fibers, paper, skin and electric components.

Measurement methods: the standard sulphur dioxide analysis method is the one pointed out in ISO/FDIS 10498 (standard project) named "Aer înconjurător - determinarea dioxidului de sulf" ("Surrounding air – sulphur dioxide determination") - UV fluorescence method.

- **Ozone**

General features: very oxidant, very reactive gas, with chocking smell. It is concentrated in the stratosphere and assures our protection against the UV radiation, which is damaging for life. The ozone present on the soil level acts as a component of the "photochemical smog". It appears following a reaction that involves mainly nitrogen oxides and volatile organic compounds.

Effects on health. The ozone concentration at ground level causes respiratory system and eye irritation. High ozone concentrations can trigger a reduction of the respiratory function.

Effects on the environment: It is responsible for certain damages caused to the vegetation through the atrophy of certain tree species in the urban areas.

Measurement methods. The standard methods for the ozone analysis and for the calibration of the ozone-related tools are:

- analysis method: UV photometric method (ISO 13964);
- calibration method: UV reference photometer (ISO 13964, VDI 2468, B1.6).

- **Nitrogen oxides**

General features. Nitrogen oxides represent a group of very reactive gases, containing nitrogen and oxygen in variable quantities. Most of these gases have no color and no smell.

The main nitrogen oxides are:

- nitrogen monoxide (NO), a colorless and odorless gas;
- nitrogen dioxide (NO_2), a brown-reddish gas, with a strong, choking smell.

Combined with the air particles, nitrogen dioxide can form a brown-reddish layer. In the presence of solar light, nitrogen oxides can react as well with hydrocarbons forming photochemical oxidants. Nitrogen oxides are responsible for acid rains, which affect the terrestrial surface and the aquatic ecosystem.

Anthropic sources: they appear in the combustion process, when fuels are burnt at high temperatures, but most often they are the result of road traffic, industrial activities, and electric energy production. Nitrogen oxides are responsible for: smog and acid rains formation, water quality deterioration, greenhouse effect, and reduced visibility in the urban areas.

Effects on people's health. Nitrogen dioxide is known as a very toxic gas both for people and for animals (its degree of toxicity is four times higher than that of the nitrogen monoxide). Being exposed to high concentrations can be fatal, while low concentrations affect the pulmonary tissue. The population exposed to this type of pollutants can experience respiratory difficulties, respiratory irritations, and pulmonary dysfunction. A durable exposure to a low concentration can destroy the pulmonary tissues, leading to pulmonary emphysema. The most affected people through the exposure to this pollutant are children.

Effects on plants and animals. The exposure to this pollutant produces serious damage to the vegetation, by whitening or destroying the plants' tissues, and reducing their growth rhythm. The exposure to nitrogen oxides can cause pulmonary diseases with animals (resembling pulmonary emphysema), while the exposure to nitrogen dioxide can reduce the animals' immunity, causing diseases such as pneumonia and flu.

Other effects. Nitrogen oxides contribute to the formation of acid rains and favor nitrate storage into the soil, which can alter the ecological balance of the environment. At the same time, they can cause tissue deterioration, paints discoloring and metal degradation.

Measurement methods: the standard method for nitrogen dioxide and nitrogen oxides analysis is mentioned in ISO 7996/1985 - "Aer înconjurător - determinarea concentraţiei masive de oxizi de azot" ("Surrounding air – the determination of massive nitrogen oxides concentration"), being chemiluminescence.

- **Lead and other toxic metals: Pb, Cd, As and Hg**

General features. Toxic metals come from coal, fuel, domestic waste combustion etc. and from certain industrial procedures. They are generally found as particles (except for mercury which is gaseous). Metals are stored in the body and trigger short and/or long term toxic effects. In case of exposure to high concentrations they can affect the nervous system, and the renal, hepatic and respiratory functions.

Measurement methods:

- the standard method for lead sampling is the same as the sampling method for PM_{10};
- the standard method for lead analysis is the one mentioned in ISO 9855/1993 "Aer înconjurător - determinarea conţinutului de plumb din aerosolii colectaţi pe filter" ("Surrounding air – determination of particulate lead content from aerosols collected on filter");
- method - atomic absorption spectroscopy.
- the standard method for measuring the concentrations of arsenic, cadmium and nickel in the surrounding air is about to be standardized by the European Committee for Standardization (CEN) and relies on manual sampling of the PM_{10} fraction (described by the EN 12341 standard).

- **Heavy metals**

In Dâmboviţa County, including the area of Târgovişte municipality and its surroundings, heavy metal emissions come from: combustion of gaseous fuels, road traffic (to a large extent) (a special role going to the use of fuels with lead derivatives as additives), metallurgical industry activities, building materials industry, and burning of dangerous (hospital) wastes (to a lesser extent).

The determinations for lead, arsenic, cadmium and nickel were carried out starting from breathable dusts - PM_{10} fraction -, and for chromium from total dusts in suspension. They were carried out using the automated DB1 and DB2 stations and the manual sampling station situated on the industrial platform in the southwest of Târgovişte municipality; for chromium, other determinations were also carried out in the locality of Doiceşti (Table 1 and Fig. 3-6).

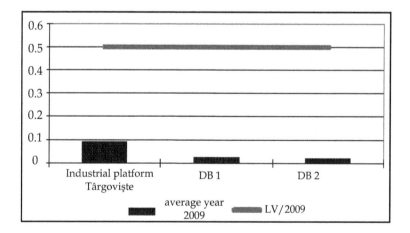

Fig. 3. Average annual concentrations: lead in the year 2009 ($\mu g/m^3$).

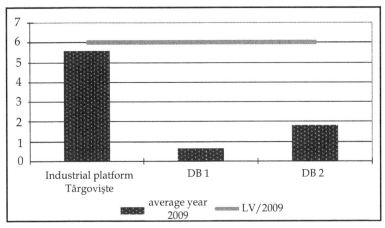

Fig. 4. Average annual concentrations: arsenic in the year 2009 (ng/m³).

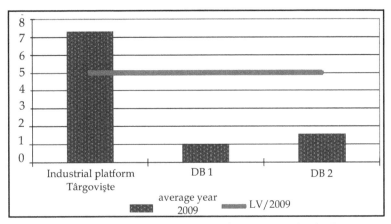

Fig. 5. Average annual concentrations: cadmium in the year 2009 (ng/m³).

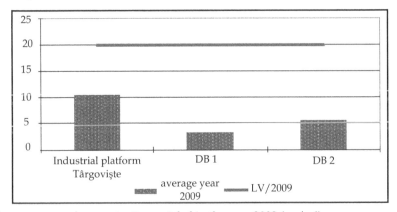

Fig. 6. Average annual concentrations: nickel in the year 2009 (ng/m³).

Station	No. of valid data	Data collecting (%)	Average	Maximum	Median
LEAD (µg/m³)					
DB1	285	78.08	0.0246	0.1623	0.0115
Industrial platform	105	28.77	0.0916	0.5712	0.0628
DB2	307	84.11	0.0205	0.0867	0.0140
ARSENIC (ng/m³)					
DB1	232	63.56	0.6237	15.9719	0.0888
Industrial platform	89	24.38	5.5645	49.9795	1.7821
DB2	251	68.77	1.7854	34.0692	0.0972
CADMIUM (ng/m³)					
DB1	232	63.56	0.9870	10.2118	0.1801
Industrial platform	89	24.38	7.2788	83.9098	1.7728
DB2	251	68.77	1.5757	9.7392	0.9380
NICKEL (ng/m³)					
DB1	232	6356	3.2938	29.5864	0.2516
Industrial platform	89	24.38	10.5071	119.8768	0.0000
DB2	235	64.38	5.4799	51.4799	2.2764
CHROMIUM (mg/m³)					
Doiceşti	193	52.88	0.000033	0.000091	0.000029

Table 1. Values for heavy metals in the year 2009 (source: AEP Dâmboviţa).

One can notice an overpassing of the yearly limit value for the protection of human health at the manual station situated on the industrial platform of Târgovişte municipality, for the indicator cadmium, under the reserve that in the year 2009 the data collecting was of 24.4% (89 measurements). The other indicators monitored did not go over the annual limit value (lead, arsenic, zinc) or the maximum admitted daily concentration (chromium). The main polluting units are: SC Cromsteel SA and SC Mechel SA from Târgovişte municipality.

- **The tropospheric ozone and other photochemical oxidants**

The inventory, according to the order 524/2000, of the sources and of the emissions highlights the following pollutants accumulation concerning the category of volatile organic compounds (VOCs) emitted in the atmosphere during the years 2007-2009 (Table 2):

Year	Total emissions in Dâmboviţa County (t/year)		
	Total VOCs	NMVOCs (non-methane volatile organic compounds)	CH₄ (methane)
2007	89127.67	14919.671	74208
2008	19177.99	8451.86	10726.13
2009	15739.2	4075.69	11663.48

Table 2. Tropospheric ozone emissions and other photochemical oxidants (source: AEP Dâmboviţa).

So, at present, one can notice a decrease of the non-methane volatile organic compounds, more precisely by 72.7% in 2009 compared to 2007, while concerning methane, a quite significant increase of the quantities emitted during the same periods was recorded, especially because of the increase of the number of cars and of the road traffic intensification, especially in the urban area (Murărescu & Pehoiu, 2009; Pehoiu, 2006).

3.2 Stationary air quality monitoring points

• **Pollution with dusts in suspension**

As we have mentioned before, in the area of Târgovişte municipality there are:

-   the sampling point from Micro XII - point 1 (Constructorului Str., no. 21). Sampling time: 24 h. Indicator – dusts in suspension: fraction PM_{10}, Unit of measurement: $\mu g/m^3$; MAC – maximum allowable concentration (50 $\mu g/m^3$);
-   the sampling point from Micro XII - point 2 (High School no. 5). Sampling time: 24 h. Indicator - total dusts in suspension (TSP), Unit of measurement: mg/m^3; maximum allowable concentration - MAC (0.15 mg/m^3); AT - alert threshold (70% of the MAC = 0.105mg/m^3) - Table 3.

Maximum value	0.058
Minimum value	0.026
Average value	0.049
Number of determinations	4
No. of situations in which the maximum allowable concentration (MAC) was exceeded	0
% MAC excess	0

Table 3. Concentrations for dusts in suspension: Târgovişte, Micro XII (source: AEP Dâmboviţa.

One can notice that in point of the concentrations of dusts in suspension, at least on the level of the quarter Micro XII of the municipality, no overpassing of the MAC was determined.

-   sampling point: AEP Dâmboviţa headquarters (Ialomiţei Str., no. 1); sampling time: 24 h. Indicators – total dusts in suspension (TSP), nitrogen dioxide (NO_2), sulphur dioxide (SO_2), ammonia (NH_3), formaldehyde (CH_2O); MAC - maximum allowable concentration (0.15 mg/m^3); AT - alert threshold (70% of the MAC = 0.105mg/m^3) - Table 4.

Just as in the case of the pollution with dusts in suspension, following the four determinations carried out, it was possible to notice that there was no MAC overpassing for dusts in suspension, the values of the alert threshold overpassing being practically insignificant.

	TSP	NO$_2$	SO$_2$	NH$_3$	CH$_2$O
Measurement units	mg/m^3	mg/m^3	mg/m^3	mg/m^3	mg/m^3
Maximum value	0.054	0.0085	0.0016	0.0049	0.0020
Minimum value	0.026	0.0070	0.0009	0.0036	0.0007
Average value	0.040	0.0076	0.0014	0.0041	0.0011
No. of determinations	4	4	4	4	4
MAC	0.15	0.1	0.25	0.1	0.012
No. of MAC overpassing	0	0	0	0	0
% MAC overpassing	0	0	0	0	0
AT	0.105	0.07	0.175	0.07	0.0084
No. of AT overpassing	0	0	0	0	0
% AT overpasing	0	0	0	0	0

Table 4. Concentrations of dusts in suspension AEP Dâmbovița (source: AEP Dâmbovița).

- sampling point: Micro VI (Unirii Blvd., no. 6)

Sampling time: 24 h. Indicators - nitrogen dioxide (NO$_2$), sulphur dioxide (SO$_2$), ammonia (NH$_3$); MAC - maximum allowable concentration, AT - alert threshold (Table 5).

	NO$_2$	SO$_2$	NH$_3$
Measurement units	mg/m^3	mg/m^3	mg/m^3
Maximum value	0.0124	0.0025	0.0062
Minimum value	0.0090	0.0018	0.0029
Average value	0.0108	0.0021	0.0048
No. of determinations	4	4	4
MAC	0.1	0.25	0.1
No. of MAC overpassing	0	0	0
% MAC overpassing	0	0	0
AT	0.07	0.175	0.07
No. of AT overpassing	0	0	0
% AT overpasing	0	0	0

Table 5. Concentrations presented by AEP Dâmbovița.

The four determinations realized in the sampling point quarter Micro VI of Târgoviște municipality highlight the overpassing of the maximum allowable concentrations for all the pollutants, accompanied by the overpassing of the alert threshold, yet in very low percentages.

AEP monitored the level of the dusts in suspension, in the area of Târgovişte, by means of its three sampling stations: AEP Târgovişte (representative for road traffic), Micro XII (representative for dusts resulted from SC Mechel Târgovişte), Civic Center (representative for dusts resulted from road traffic and other sources after dispersion (SC Mechel SA, SC UPET SA).

Following four determinations, we noticed that the values of the average concentrations in 24 hours went over the MAC (0.15 mg/m^3) in all the three sampling points. The frequency of the average overpassing in 24 h for the area of Târgovişte is of 13.94%, the maximum overpassing being of 0.267 mg/m^3 (in the point AEP headquarters) compared to 0.15 mg/m^3 (MAC).

The values of the average yearly concentrations were above the annual MAC (0.075 mg/m^3) in all the three sampling points and globally in the area of Târgovişte (0.116 mg/m^3). The main reasons are the dust emissions containing iron oxides and ferrous metals (SC Mechel SA), but also the intense road traffic.

In order to observe the air quality status and the pollution in the north of Târgovişte municipality, the sampling points from the industrial area of Doiceşti were taken into account as well, having as sampling point the locality's train station, also because the wind direction determines the transport of the dusts from this location to the municipality, along the valley of Ialomiţa River. Sampling time: 24 h. Indicators - nitrogen dioxide (NO$_2$), sulphur dioxide (SO$_2$), sulphuretted hydrogen (H$_2$S), Cr^{6+}, total dusts in suspension (TSP); MAC - maximum allowable concentration; AT - alert threshold (70% of the MAC = 0.105 mg/m^3) - Table 6 and Fig. 7.

	TSP	NO$_2$	SO$_2$	H$_2$S	CrO$_3$
Measurement units	mg/m^3	mg/m^3	mg/m^3	mg/m^3	mg/m^3
Maximum value	0.057	0.0204	0.0033	0.0045	0.0001000
Minimum value	0.029	0.0087	0.0013	0.0017	0.0000840
Average value	0.037	0.0132	0.0020	0.0028	0.0000893
No. of determinations	4	4	4	4	4
MAC	0.15	0.1	0.25	0.008	0.0015
No. of MAC overpassing	0	0	0	0	0
% MAC overpassing	0	0	0	0	0
AT	0.105	0.07	0.175	0.0056	0.0010
No. of AT overpassing	0	0	0	0	0

Table 6. Concentrations reported in the point Doiceşti Trains Station (source: AEP Dâmboviţa).

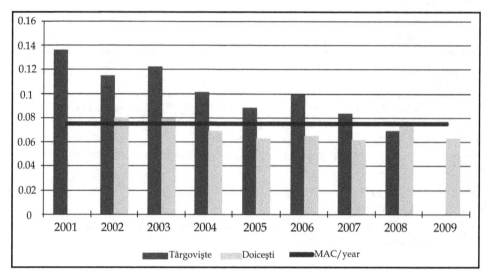

Fig. 7. Evolution of the concentrations of total dusts in suspension during the years 2001-2009 (TSP) - mg/m³

In Doiceşti locality, one monitored the level of the pollution with settleable dusts, nitrogen dioxide, sulphur dioxide, sulphuretted hydrogen, dusts in suspension, given the specific emissions from the main polluting economic agents in the area: U.E. Doiceşti, which uses black oil and brown coal as fuel to generate electricity and SC Nubiola România SRL (a company producing whiteners and green chrome oxide) - the SO_2 and possibly a sulfuretted hydrogen source in the process of sulphur combustion needed to obtain ultramarine.

Other potential pollutants are SC. Soceram SA (producer of bricks, ceramic materials) and road traffic. The evolution of the concentrations of total dusts in suspension, during the period 2001 - 2009, clearly highlights the diminution in the atmosphere of these polluting emissions, both for the municipality and for Doiceşti locality.

- **Pollution with settleable dusts**

AEP supervises the level of the settleable dusts in the areas of Târgovişte and Doiceşti. Under exceptional conditions of air masses transportation along Ialomiţa River, settleable dusts from the industrial area of Fieni can end up in Târgovişte and its surroundings. That is why we consider it necessary to include Fieni locality as well in the present study. In these areas, a characteristic feature is the pollution with dusts, the main polluting sources being:

- in Târgovişte: SC Mechel SA, SC Upet SA and to a lesser extent the road traffic; the major impact is felt under the form of dusts in suspension;
- in Doicesti: UE Doiceşti, SC Soceram SA, road traffic; given the emissions' features, the major impact is felt under the form of settleable dusts. Dusts in suspension have been monitored in the area starting with June 2002.
- in Fieni: SC Carpatcement Holding SA (producer of lime and cement); the major impact is felt at the same time under the form of settleable dusts. Dusts in suspension have been monitored in the area starting with March 2003.

Concerning each locality, the average annual quantities of settleable dusts did not go over the annual MAC (204 t/km²/year) in none of them, yet in different sampling points the annual average values have been exceeded, namely in the points situated in inhabited areas in which the impact of the dusts emissions from the sources is high (Fig. 8).

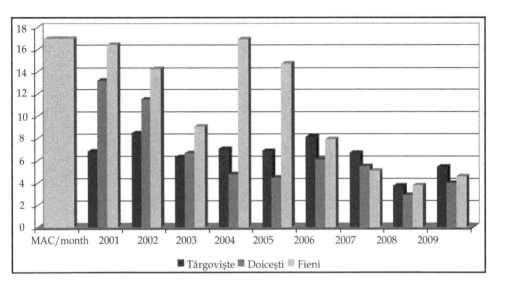

Fig. 8. Evolution of the concentrations of settleable dusts during the years 2001-2009 (mg/m³).

3.3 Experimental results and syntheses concerning the automatic monitoring of the emissions of atmospheric pollutants

- Târgovişte area

For the indicator dusts in suspension - the PM_{10} fraction - (Procor headquarters, industrial platform), on 13 occasions one recorded values above the limit value (50 µg/m³) in 24 hours (out of the 15 measurements carried out). In the year 2009, the level of dusts in suspension was analyzed in the localities Târgovişte (PM_{10} - manual station located on the industrial platform and industrial automatic station - DB1), Doiceşti (total dusts in suspension (TSP) - manual station) and Fieni (industrial automatic station - DB2) - Table 7 and Fig. 9-13.

Station	No. of valid data	Data collecting (%)	No. of data >LV	Frequency of the overpassing (%)	Average ($\mu g/m^3$)	Maximum ($\mu g/m^3$)	Median ($\mu g/m^3$)	Percentile 98 ($\mu g/m^3$)
TÂRGOVIŞTE								
DB1, automatic	272	74.52	6	2.21	19.665	55.373	15.634	48.942
DB1, gravimetric	286	78.36	1	0.35	19.594	50.150	19.079	43.848
Industrial platform, gravimetric	142	38.90	87	61.27	57.629	118.387	57.364	106.567
DOICEŞTI								
Train Station	194	53.15	0	0	0.063	0.148	0.059	0.136
FIENI								
DB2, automatic	335	91.78	11	3.28	22.101	89.920	19.425	55.581
DB2, gravimetric	307	84.11	8	2.61	18.739	75.043	15.808	53.060

Table 7. Indicators PM_{10} per 24 h in 2009 (source: AEP Dâmbovița).

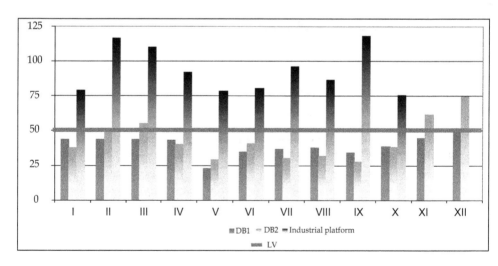

Fig. 9. Maximums/24h of the PM_{10} indicator, measured gravimetrically (months) - year 2009, compared to LV/24h ($\mu g/m^3$).

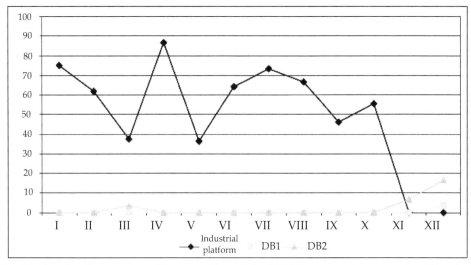

Fig. 10. Monthly frequencies of the exceeding of the LV/24h – dusts in suspension (PM_{10}) in the year 2009 (%).

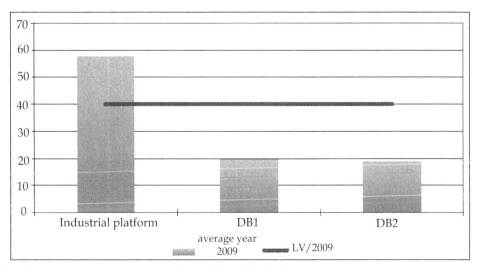

Fig. 11. Average annual concentrations of the PM_{10} indicator, measured gravimetrically ($\mu g/m^3$).

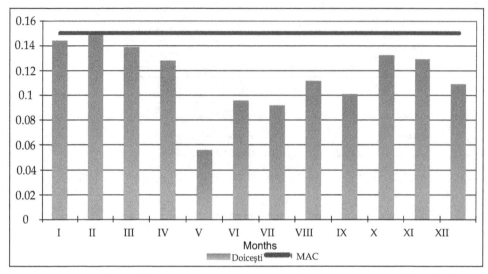

Fig. 12. Maximums/24h of the TSP indicator in Doiceşti Commune - year 2009 (µg/m³).

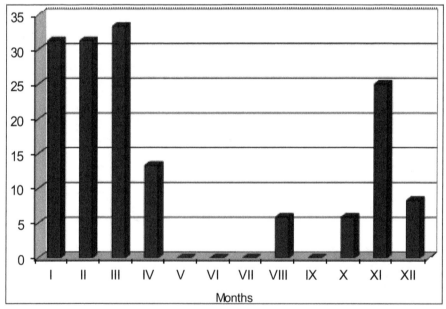

Fig. 13. Monthly frequencies for the exceeding of the AT/24h, Doiceşti locality, TSP indicator (%), year 2009.

According to the values recorded for the PM_{10} indicator, in April 2009, one can notice the increase of the frequency for the exceeding of the limit value to 86.7% (compared to 37.5%, the previous month), the average monthly concentration being of 68.7 µg/ m³, compared to 52.6 µg/ m³ in March the same year (Table 8).

Maximum value of the concentration (µg/m³)	Minimum value of the concentration (µg/ m³)	Average of the values recorded (µg/ m³)	Frequency of the overpassing of the LV%
92.2	34.5	68.7	86.7%

Table 8. Average, maximum, minimum monthly quantities for settleable dusts and the frequencies for the exceeding of the monthly MAC (Micro XII, point 1, Târgovişte).

At the Automatic station DB-1, Micro 11, Târgovişte, no exceeding of the limit value for the indicator PM_{10} gravimetric, the average monthly value being of 20.73 µg/ m³ (Fig. 14).

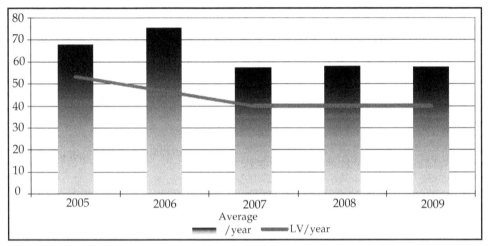

Fig. 14. Evolution of the concentrations of breathable dusts - PM_{10} fraction, Târgovişte, Micro XII, manual station (µg/ m3), years 2005-2009.

The dusts concentration in the atmosphere in the sampling area is influenced by the input from nearby sources (activities in the domain of building materials) and of the nearby sources from a distance of about 1 km (activities such as steel production, slag processing etc.). At the same time, the PM_{10} concentration can have high values depending on the evolution of the meteorological phenomena: high relative humidity (RH), atmospheric pressure and wind intensity may lead to the appearance of "peaks" of the PM_{10} concentrations, because they can favor the agglomeration of the particles.

In Târgovişte municipality, the main sources generating dust emissions (with significant impact in the area around their activity) are the activities of the economic agents from the area of the industrial platform, which operate in the metallurgical domain, in the domain of metallic ware, road transport, building materials, concrete production and road traffic.

In the case of gaseous pollutants, at the Automatic monitoring station DB-1 there was no exceeding of the limit values or the information thresholds for the average hourly or daily concentrations for the indicators monitored. The main sources emitting gaseous pollutants in the area of Târgovişte, with a potential impact on the air quality are: metallurgical companies, industrial companies that use solvents, economic agents operating in the domain of metallic ware, road transport, building materials, fuel distribution stations and road traffic.

4. Air quality status

In Târgovişte municipality one monitors daily concentrations (sampling time: 24 h) for the indicators: breathable dusts, respectively the fraction with the diameter <10μm (PM_{10}), settleable dusts, highlighting the quantity of (settleable) dusts deposited during a 30-day interval on a 1 m^2 area, this being a characteristic indicator for highlighting the pollution with heavy particles in suspension, which are later on deposited on the ground.

At the same time, the automatic station DB-1 Târgovişte, which continually monitors meteorological parameters (temperature, wind speed, wind direction, solar radiation intensity, precipitations quantity, atmospheric pressure), gaseous pollutants (nitrogen oxides, sulphur dioxide, carbon monoxide, tropospheric ozone) and dusts in suspension (breathable - PM_{10} fraction) transmits the data, in real time, to the panels dedicated to public information (external panel - Prefecture Plateau of Târgovişte - and internal panel - AEP Dâmboviţa headquarters) - Fig. 15.a-b.

In Doiceşti locality, air quality is monitored by supervising the indicators total dusts in suspension, settleable dusts and gaseous pollutants (sampling point: manual/urban station).

4.1 Examples of chronological series recorded during a day, (April 1-2 and 18, 2009), at the automatic station DB-1 Târgovişte (concentration in μg/m³)

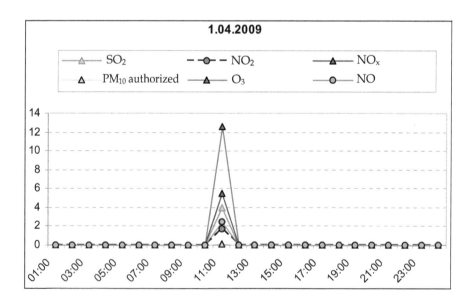

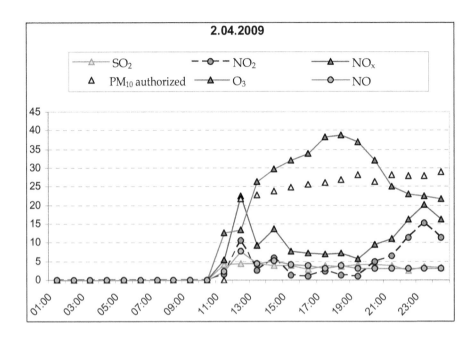

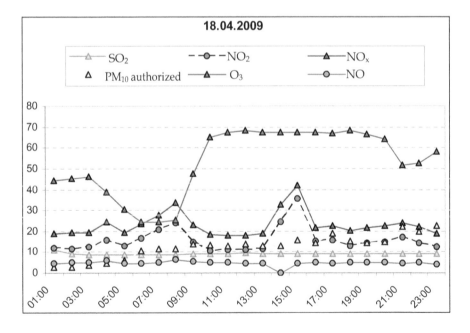

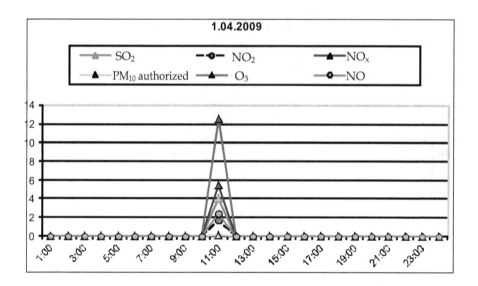

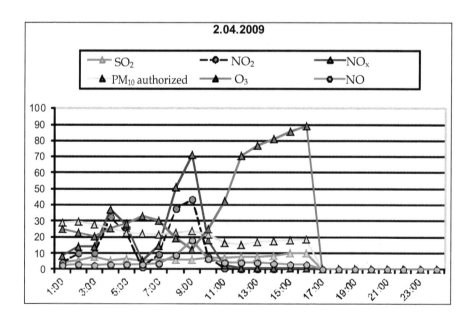

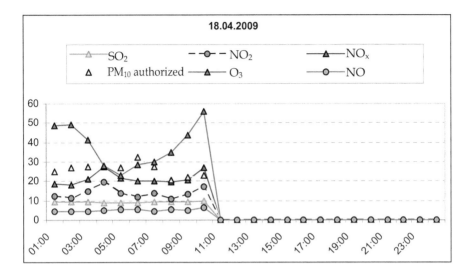

5. Air quality management

5.1 Goals and measures for air quality management

Air pollution represents a serious problem, with short, medium and long term effects. The air, as environmental factor, is submitted to an intense local pollution, especially in the urban area under analysis, because of the industrial activity, because of the intensification of the road traffic and also because of the burning of domestic waste. The air pollution effects can be direct, affecting the local population's health condition, and also indirect, affecting fauna, flora and building materials.

The goals and measures foreseen on the level of Dâmbovița County and implicitly for all the localities under analysis, in order to improve the air quality in the urban area, in agreement with the quality standards, refer to: reducing the impact of the road traffic on the air quality, reducing the emissions from individual heating systems, assuring the control of the emissions from industrial sources and installing de-pollution equipments for all the sources with a major impact.

There are priorities concerning volatile organic compounds (VOC) emissions control; they aim to limit the emissions coming from the use of organic solvents in certain activities and installations (protective cover, dry cleaning, fabrication of covering compounds and varnishes; surface cleaning and degreasing; wood impregnation) and to meet the legal regulations in this domain.

Considering the fact that industrial pollution represents the main source of pollution for all the environmental factors (affecting the quality of the air, water, soil, generating different types of waste and using natural resources and energy), the purpose of the *integrated environmental protection system* is the implementation of preventive measures or the emissions reduction with the purpose of protecting the environment as a whole.

The industrial branch with the highest impact on the environmental factors is metallurgical industry, the air quality being affected by: emissions resulted from raw matter preparation (iron melting), final products processing, raw matter and auxiliary products transport and storage.

A significant impact on the environmental factors comes from the building materials industry as well (through the existence of the previously mentioned economic agents); their specific activities determine the elimination of large quantities of dusts and green house effect gases in the atmosphere.

On the level of Dâmbovița County, eleven environmental authorizations have been issued concerning the emissions of green house effect gases, for installations under the regulations of the EU-ETS (EU - Emission Trading System) Directive, for the period 2008-2012 (The directive 87/2003/CE was adopted by the Romanian legislation through the governmental decision H.G. no.780/2006)[1].

The integrated authorizations contain demands and limit values regarding the emissions, which try to make sure that all the adequate preventive measures for the environmental protection have been taken. The limit values concerning emissions are established based on the best available techniques. In order to support the EU member states in the application of the IPPC (International Plant Protection Convention) Directive, among the member states and the afferent industry, there has been an information exchange concerning the BAT - Best Available Techniques. The purpose of this information exchange is to balance differences on a technological level in the European Community and to promote universal limit values and techniques.

In the areas affected by industrial pollution, through adjustment programs annexed to the environmental authorizations emitted for the polluting economic agents, one established measures for the reduction of the industrial activities' impact on the environmental quality. They are found as well in the Local Action Plan for the Environement of Dâmbovița County (Planul Local de Acțiune pentru Mediu al Județului Dâmbovița), reviewed even since the year 2007. Among the local economic agents concerned there are: SC Mechel SA, SC Termica S.A, SC Oțelinox SA, all from Târgoviște, SC Termoelectrica SA București – the branch of Doicești, SC Soceram SA București - the branch of Doicești, SC Carmeuse Holding SRL – working point of Fieni, SC Carpatcement Holding SA - the branch of Fieni.

Following the approval by the National Agency for Environmental Protection (Agenția Națională pentru Protecția Mediului - ANPM) of the Propositions of measure plans for

---

[1]A certificate concerning the green house effect gases emissions represents the title giving an installation the right to emit a ton of carbon dioxide equivalent during a definite period; it is valid for meeting the goal of the governmental decision H.G no. 780/2006 and transferable under the conditions foreseen by this normative act.

green house effect gases emissions monitoring and reporting for the year 2010, the environmental authorizations concerning this type of emissions were reviewed.

The joint implementation (JI), according to the Kyoto Protocol, concerning the advantages of the project for Târgovişte municipality, considered the following aspects:

- producing non-polluting electric and thermal energy using modern, reliable installations, with high energetic efficiency;
- assuring the necessary thermal energy for the municipality of Târgovişte from its own source for the following 20 years;
- realizing 1,959,461 Euro worth investments in the infrastructure without using funds from the state budget;
- creating new jobs;
- encouraging other foreign investments in the municipality.

The ERU 04/04 Contract was concluded between Senter Novem Agency from Holland, from the part of the Government of Holland, and SC Nuon Energy Romania SRL Sibiu, Romania - Project Developer, on September 28, 2004. In the year 2010 a transfer agreement was concluded on the basis of the ERU 04/40 Contract from SC Nuon Energy Romania SRL to SC Termica SA Târgovişte. The emission reductions generated by the project beginning with January 1, 2010 are dealt with by SC Termica SA Târgovişte and the above-mentioned Hollandese agency (Local Plan of Action for Environment of Dâmboviţa County, 2010).

6. The air quality in relation to the population's health condition

The environment in which people live is first of all defined by the quality of the air, of the water, of the soil, of the dwellings, of the foods they eat, and of the environment in which they deploy their activity. Closely connected to these factors, influenced or determined immediately or after a certain period of time is the population's health condition.

The evaluation of the population's health condition consists in the identification of the hazard factors that in the urban area have an influence on: air quality; potable water supply; gathering and removing liquid and solid residues of any nature; urban noise; habitat – improper conditions (noise, light, population agglomeration etc.); services quality (of all types) provided to the population (Pehoiu et al., 2006).

Knowing and determining some environmental hazard factors is particularly important and may constitute one of the most valuable activities related to the promotion and maintaining of the population's health condition (Pehoiu &n Murărescu, 2009).

The action of the environmental factors on human health is very diverse. When the pollution intensity is higher, the action on organisms is immediate. However, more often than not, their action has a low intensity, determining a chronic, long-term action, the quantification of the effect becoming difficult to achieve. The atmospheric pollution in general and especially the exposure to dusts in suspension produces diseases of the respiratory, digestive, osteo-muscular and nervous systems and of the sensory organs, affecting all the age groups; however, when it comes to children, it determines an early predisposition to respiratory diseases and bronchial asthma. The irritating capacity of the dusts in suspension increases when, in the air, there are other irritating respiratory pollutants, such as SO_2 and NO_2, as a synergic effect comes into operation from the SO_2 - dusts in suspension and NO_2 - dusts in suspension (Table 9).

Types of diseases	Number of occurrences
1. Diseases of the respiratory system, of which: - acute bronchitis - chronic bronchitis - bronchial asthma	 1963 123 110
2. Digestive diseases	8168
3. Cardiovascular diseases	4841
4. Endocrine and metabolic diseases	2966
5. Malignant tumors, of which: - pulmonary cancer	 21
6. Congenital malformations	16
7. Renal system diseases	7962
8. Flu	43

Table 9. Specific morbidity on the level of the year 2009 in the urban area of Dâmbovița County (source: Public Health Direction of Dâmbovița County).

In the prevention of diseases caused by the exposure of the population to different atmospheric pollutants, a special importance goes to their prophylaxis. In this sense, one should have in view the maintaining of the concentration of the toxic substances from the environment under the level of the maximum allowable concentrations (MAC) stipulated in the norms (STAS).

7. Conclusions

In Dâmbovița County and implicitly in Târgoviște municipality and its surroundings, the distribution of the sampling equipments available and the organization of the air quality monitoring network was carried out in the sense of assuring especially the monitoring of the areas most likely to be affected by impact pollution.

The pollutants to be monitored, the methods of measurement, the limit values, the alert and the information thresholds and the criteria for the location of the monitoring points are established by the national legislation concerning the atmospheric protection and meet the requirements foreseen by the European regulations.

The air quality has undergone a slight improvement compared to the previous years from the viewpoint of the pollution with different dusts.

The gaseous pollutants (nitrogen dioxide, sulphur dioxide, ammonia, sulphuretted hydrogen, formaldehyde, oxidant substances, and carbon monoxide) are not present in the atmosphere in concentrations above the allowable limits in the areas in which they are monitored.

The dominant pollution is represented by dusts in suspension (in the area of Târgovişte, in Doiceşti and its surroundings), and by pollution with settleable dusts in Fieni area.

8. References

Agenţia Naţională pentru Protecţia Mediului (National Agency for Environment Protection), statistical data, 2000-2010, Bucharest, Romania.

Anuarul statistic al judeţului Dâmboviţa (Statistical Yearbook of Dâmboviţa County), 2000-2010, Târgovişte, Romania.

Direcţia de Sănătate Publică Dâmboviţa (Public Health Department Dâmboviţa), statistical data, 2000-2010, Târgovişte, Romania.

Direcţia Judeţeană de Statistică Dâmboviţa (Statistical Department of Dâmboviţa County), statistical data, 2000-2010, Târgovişte, Romania.

Ministerul Mediului şi Pădurilor (Ministry of Environment and Forestry), statistical data, 2000-2010, Târgovişte, Romania.

Murărescu, O.; Pehoiu, G. (2009). Integrated management of environmental pollution due tu the County Dâmboviţa industrial activiy. *Annals - Food Science and Technology*, Vol. 10, Issue 2, 2009, pp. 681-686, ISSN 2065-2828, "Valahia" University, Târgovişte, Romania.

Pehoiu, G. (2003). *Câmpia Înaltă a Târgoviştei. Studiu de geografie umană şi economică* (Târgovişte High Plain. Study of Human and Economical Geography), Cetatea de Scaun, ISBN 973-7925-01-7, Târgovişte, Romania.

Pehoiu, G. (2006). Actual aspects related to the quality of the air in the county Dâmboviţa, in report with the status of health of the population. International Multidisciplinary Symposium „Universitaira Sempro 2006", *Ingineria mediului* (Environment Engineering), ISSN: 1842-4449, Petroşani, 2006.

Pehoiu, G.; Muică, C.; Sencovici, M. (2006). Geografia *mediului cu elemente de ecologie* (Environment Geography with Elements of Ecology), Transversal, ISBN (10) 973-7798-32-5, ISBN: (13) 978-973-7798-32-9, Târgovişte, Romania.

Pehoiu G.; Murărescu O. (2009). Climate Change Impact on Environment and Health of the Population in Dâmboviţa County, Romania, *Proceedings of the 2nd WSEAS International Conference on Climate Changes, Global Warming, Biological Problems* (CGB '09), ISSN:1790-5095, ISBN: 978-960-474-136-6, Morgan State University, Baltimore, USA, November 7-9, 2009.

Pehoiu G.; Simion T.; Murărescu O.M. (2005). Climat urbain, et la pollution de l'air dans les villes industrielles. Etude de cas - La ville de Târgovişte (Roumanie). XVIIIe Colloque Internationale de Climatologie, Genova, 7-11 septembrie 2005. In: *Climat urbain, ville et architecture*, ed. Gerardo Brancucci, pp. 39-42, Laboratoire de Géomorphologie Apliquée, Département Polis, Université de Gênes, Retrieved from http://climato.ulg.ac.be/doc/AIC-table.pdf.

Pehoiu G. (2008). The impact of human activities on environmental quality in Dâmbovița county. In *Present Environment and Sustainable Development*, Vol. 2, pp. 283-296, „Alexandru Ioan Cuza" University, ISSN 1843-5971, Iași, Romania.

Planul Local de Acțiune pentru Mediu al Județului Dâmbovița (Local Plan of Action for Environment of Dâmbovița County), 2010, Târgoviște, Romania.

Agenția pentru Protecția Mediului - Dâmbovița (Agency for Environmental Protection - Dâmbovița County), *Raport privind starea mediului în județul Dâmbovița* (Report regarding state of the environment in Dâmbovița County), 2000-2010, Târgoviște, Romania.

2

Effect of Air Pollutants on Vegetation in Tropical Climate: A Case Study of Delhi City

Sumanth Chinthala and Mukesh Khare
Indian Institute of Technology Delhi
India

1. Introduction

Urban air pollution is a serious problem in both developed and developing countries (Li, 2003). As a rapidly expanding centre of government, trade, commerce and industry, Delhi, the Indian capital, has been facing many air pollution related problems. The vehicular exhaust contributes significantly to the pollution load in Delhi. The plant species are severely affected by various pollutants emitted from different sources e.g. SO_x, NO_x and particulates. To maintain ecological balance in this fastly developing capital, there is an immediate necessity to assess the effect of the air pollutants on the plant speices so that the strategies can be formulated and implemented to protect the species. Plants remove air pollutants by three mechanisms: absorption by the leaves, deposition of particulates and aerosols over leaf surfaces, and fallout of particulates on the leeward side of the vegetation (Tewari, 1994; Rawat and Banerjee, 1996). As a result, the chlorophyll concentrations in the leaf which is responsible for the photosynthetic activity may decrease (Seyyednejad et al, 2011). Hence, the ability of the plant to tolerate the air pollution gets affected. Plantation of tolerant tree species will have a marked effect on varied aspects of the quality of the urban environment and the cleanliness of life in a city (Bamnia et al, 2011). The Anticipated Performance Index (API) has been used as an indicator to assess the capability of some of the predominant species present in Delhi. Further, the air pollution tolerance index (APTI) of the plants needs to be monitored and checked for the predominant species that are located in the city.

2. Topography of Delhi

Delhi is positioned with the Great Indian Desert (Thar Desert) of Rajasthan to the west and southwest, central hot plains to the south and gangetic plains of Uttar Pradesh/Uttaranchal to the east while cooler hilly regions to the north. It is situated at latitude 28°24'17" and 28°53'00" North; Longitude 76°45'30" and 77°21'30" (East) at elevation of 216 m above the mean sea level (msl). Delhi's climate is very hot in summer (April - July) and cold in winter (December - January). The average temperature can vary from 25° C to 45° C during the summer and 22° C to 5° C during the winter respectively. The topography of the city is manly urban plain having two main features - the Ridge and the river Yamuna. The Ridge refers to an area inhabited by extraordinary plants and fierce animals.

3. Air quality status in Delhi

The Air quality in Delhi is primarily affected by the vehicular exhausts, small scale industries, power plants and biomass burning. The relative contribution of various sources for the air pollution in Delhi has been shown (Table 1&2). It can be found that the vehicular exhausts contribute a significant amount to the air pollution in the city. The pollutants include SPM, RSPM, SOx, NOx and Benzene. It has also been reported that the benzene concentration has exceeded seven times that of the permissible limit during the winter season. In summers, the concentration of the pollutant has been below the permissible standards prescribed. A Source apportionment study to identify the sources of the benzene is to be carried out (DPCC, 2011).

Sl No	Sector	Percentage of emissions
1	Roads (Paved and Unpaved)	52-53
2	Area sources	18-20
3	Vehicles	3-5%
4	Industrial	20-22

Table 1. Prominence of Sources of PM_{10} for Delhi city (Source: CPCB, 2010).

Sl No	NOx		SOx	
	Sector	Percentage of emissions	Sector	Percentage of emissions
1	Industrial	78-80	Industrial	98
2	Vehicular	18-20	Vehicular	1
3	Area Sources	2-3	Area Sources	1

Table 2. Prominence of Sources of NOx and SOx for Delhi city (Source: CPCB,2010).

4. Vegetation covers in the city

The landscape of Delhi consists of a broad spectrum of environments ranging from the city forests to highly modified artificial landscapes in certain parks (Khera et al, 2009). Among the 260 different varaities of species present in delhi, only 42 species are native species. The most dominant speicies is *Prosporis Juliflora*, a tree from central America that was introduced around 1915 in order to afforest the central ridge forests. The other commonly found tree species in Delhi include, Eucalyptus sp., Ficus benghalensis, Ficus religiosa, Mangifera indica, Melia azedarach, and Syzygium jambolanum, Alstonia scholaris, Azadirachta indica, and Cassia fistula. Since the majority of the tree species are not a native varaities, their ability to sustain and sequestrate the air pollutants needs to be evaluated at regular intervals. The features of the predominant tree species has been shown in the table 3.

The effect of air pollutants is showing a significant effect on the vegetation in the city. The predominant species, *Prosporis Juliflora* in the capital's central ridge has been considered as most sensitive to air pollution (Seyyednjad et al, 2011). With the presence of the most sensitive species in a vast area in the heart of a city, the amount of pollutants absorbed by the vegetation may get reduced. Further, some of the tree species may emit VOC's which play an important role in the atmospheric chemistry affecting the local air quality.

Sl no	Name	Scientific name	Family
1	Shoe Babool	Leacena leucophloea	Mimoseae
2	Gulmohar	Delonix regiosa	Caesalpiniaceae
3	Neem	Azardirachta indica	Meliaceae
4	Vilayati kikkar	Acacia farnesiana	Mimoseae
5	Bankayan	Melia azedarch	Meliaceae
6	Bottle brush	Calistemon citrinus	Myrtaceae
7	Torch Tree	Ixora parviflora	Rubiaceae
8	Chikkoo	Archise sapota	Sapotaceae
9	Arjun	Terminalia arjuna	Combrataceae
10	Babool	Accacia nelotica	Mimoseae
11	Shesham	Delbergia sisso	Papilionaceae
12	Jangal Badam	Terminalia catappal	Combrataceae
13	Ashoka	Polyalthia longifolia	Annoniaceae
14	Kanju, Papadi	Holiptelia integrifolia	Ulmaceae
15	Amrood	Psidium guyaua	Myrtaceae
16	Mahua	Madhuca indica	Sapotaceae
17	Mulberry	Morus alba	Moraceae
18	Semal	Bombax ceiba	Bombaceae
19	Popular	Populus trimuloides	Siliaceae
20	Plums	Prunus Comminis	Rosaceae
21	Lemon	Citrus lemon	Rutaceae
Sl no	Name	Scientific name	Family
22	Kanchnar	Bauhinia Variegata	Caesalpiniaceae
23	Bel	Aegle marmelos	Rubiaceae
24	Satni	Alstonia Scholaris	Apocynaceae
25	Mango	Magnifera Indica	Anacardiaceae
26	Blue gum	Eucalyptus globulus	Myrtaceae
27	Jamun	Syzygium cuminii	Myrtaceae
28	Peepal	Ficus religiosa	Moraceae
29	Teak	Tectona grandis	Verbenaceae
30	Kadam	Anthosephalus cadamba	Rubiaceae
31	Wolly Mopming Glorry	Argyreia roxburghira	Caesalpiniaceae
32	Amaltas	Cassia fistula	Caesalpiniaceae
33	Banyan tree	Ficus bengalensis	Moraceae
34	Indian rubber	Ficus elastica	Moraceae

Table 3. Predominant Tree species in Delhi (CPCB, 2007).

Regional as well as global scale VOC emission, from vegetation, may dominate over anthropogenic sources of emission (Guenther et al., 1995). Out of the nine commonly occurring tree species in Delhi, VOC emissions were found in six species, namely, *Eucalyptus sp.*, *Ficus benghalensis*, *Ficus religiosa*, *Mangifera indica*, *Melia azedarch*, and *Syzygium jambolanum*, at higher concentrations. Further it has also been observed that that these emissions are dependent up on sunlight and temperature (Tingey et al., 1979; Lamb et al., 1987; Padhy and Varsheney, 2005).

5. Quantifying the effects of air pollution using APTI and API

The effect of air pollution on the plants can be quantified using a parameter, air pollution tolerence index (APTI) (Singh and Rao, 1993). The APTI is a function of total chlorophyll content of the leaf, pH, relative water content and the ascorbic acid content. The APTI for a particular tree species is given in eq (1)

$$APTI = (A \, (T+P) + R)/10 \tag{1}$$

Where A= Ascorbic acid content in mg/g dry weight,
T= Total Chlorophyll content in mg/g dry weight,
P= pH of the Leaf extract,
R = Relative water content (%)

The plant species can be convinently grouped based on the APTI values (Table 4).

Sl no	APTI	Response
1	30 -100	Tolerant
2	29-17	Intermediate
3	16-1	Sensitive
4	<1	Very sensitive

Table 4. Grouping of plants using APTI (Source: Lakshmi et al, 2006).

Further, the parameters of the APTI have been correlated with the increasing concentration of the pollutants. It has been found that the relative water content , total chlorophyll are negatively correlated and ascorbic acid is positively correlated with all the pollutants. The pH values are negatively correlated in the case of NO_x (Table 5).

	Relative water content	pH	Total chlorophyll	Ascorbic acid
SPM	Negative	Positive	Negative	Positive
RSPM	Negative	Positive	Negative	Positive
SO_2	Negative	Positive	Negative	Positive
NO_2	Negative	Negative	Negative	Positive

Table 5. Correlation between Air pollutants and Biochemical parameters (Govindaraju et al, 2011).

The APTI index shows the effect of the pollutants only on the biochemical parameters. In order to combat air pollution by planning the green belt development in a particular area, many socio-economic factors are to be considered. Hence the anticipated performance index has been used to determine the same (Govindaraju et al, 2011). The method contains a grading system where a tree species is graded based on various parameters. The API has been calculated for predominant tree species in Delhi. The table 6 shows the parameters used to grade the performance of a particular tree species. It includes the parameters like APTI along with the socio-economic parameters. Based on the current grading system, a tree can secure a maximum of 16 positive points. These points were scaled to a percentage system and based on the score obtained, the category has been assessed. Table 7 shows the assessment categories along with the scores.

Grading Character	Parameter	Pattern of assessment	Grade alloted
Tolerence	APTI	9.0-12.0	+
		12.1-15.0	++
		15.1-18.0	+++
		18.1-21.0	++++
		21.1 -24.0	+++++
Biological and Socio -economic	Plant habit	Small	-
		Medium	+
		Large	++
	Canopy Structure	Sparse /Irregular/ Globular	-
		Spreading crown/open/ semi dense	+
		Spreading dense	++
	Type of Plant	Deciduous	-
		Evergreen	+
Laminar structure	Leaf Size	Small	-
		Medium	+
		Large	++
	Texture	Smooth	-
		Coriacious	+
	Hardness	Delineate	-
		Hardy	+
	Economic Value	Less than Three uses	-
		Three or Four uses	+
		Five or more uses	++

Table 6. Gradation of plant species based on APTI as well as morphological parameters and socio – economic importance (Govindaraju et al, 2011).

Grade	Score (%)	Assesment category
0	Up to 30	Not recommended
1	31-40	Very poor
2	41-50	Poor
3	51-60	Moderate
4	61-70	Good
5	71-80	Very good
6	81-90	Excellent
7	91-100	Best

Table 7. Anticipated performance Index (API) for species (Govindaraju et al, 2011).

To determine the effect of air pollutants on the tree species in Delhi, the API values have been calculated. Since the data on the APTI values of the tree species considered are not available, the analysis is performed for both the extreme cases i.e for maximum APTI and for the minimum APTI. Table 8 and 9 shows the maximum and minimum values of the API obtained.

It can be shown from the table 9 that even though the APTI of a species is negligible (less than 9) , but still the API of a plant species are securing a reasonable grades. Out of the eight species, three are *good*, two are *moderate*, two are *poor* and one is *very poor*.

Sl	Species	a	b	c	d	e	f	g	h	Total	(%)	Category
1	Prosporis juliflora	+	+	+	-	-	++	+	+++++	11	68.7	Good
2	Eucalyptus sp	++	+	+	++	+	++	+	+++++	15	93.7	Best
3	Ficus benghalensis	++	+	+	++	+	++	+	+++++	15	93.7	Best
4	Ficus religiosa	++	+	+	++	+	+	+	+++++	14	87.5	Excellent
5	Mangifera indica	++	+	+	++	+	++	+	+++++	15	93.7	Best
6	Alstonia scholaris	++	-	+	++	+	+	+	+++++	13	81.2	Excellent
7	Azadirachta indica	++	++	-	-	-	++	+	+++++	12	75	Very good
8	Cassia fistula	++	++	-	++	+	+	+	+++++	14	87.5	Excellent

Table 8. Maximum API for the predominant plant species in Delhi.

Sl	Species	a	b	c	d	e	f	g	Total	(%)	Category
1	Prosporis juliflora	+	+	+	-	-	++	+	6	37.5	Very poor
2	Eucalyptus sp	++	+	+	++	+	++	+	10	62.5	Good
3	Ficus benghalensis	++	+	+	++	+	++	+	10	62.5	Good
4	Ficus religiosa	++	+	+	++	+	+	+	9	56.2	Moderate
5	Mangifera indica	++	+	+	++	+	++	+	10	62.5	Good
6	Alstonia scholaris	++	-	+	++	+	+	+	8	50	Poor
7	Azadirachta indica	++	++	-	-	-	++	+	7	43.7	Poor
8	Cassia fistula	++	++	-	++	+	+	+	9	56.2	Moderate

where a= plant habitat; b= canopy structure; c= type of plant; d=leaf size; e= texture; f= economic importance; g= hardiness ; h = APTI.

Table 9. Minimum API for the plant species in Delhi.

Since the Delhi's most predominant species is showing very poor performance, it can be said that if the APTI value decreases, the API may not indicate the performance of a plant species. Since the reduction in the APTI values may also affect the other biochemical and socio-economic parameters, a modified index is proposed by adding a negative weightage. For every unit reduction in the positive value of the APTI, a negative weightage is proposed.

Further, it has been found that among all the pollutants, SPM has the maximum effect on the biochemical parameters; and NOx has the least effect. The pollutants can be arranged in the following order. SPM> RSPM> SO_x> NO_x (Govindaraju et al, 2011). To incorporate the effect of increase of the concentrations of these pollutants, an additional weightage of 1, 0.75, 0.5 and 0.25 may be added to the negative weightage, respectively. This additional weightage will be added only when the concentration of the polluants exceed the specified standards.

Sl	APTI	Weightage	SPM	RSPM	SO_x	NO_x
1	+++++	0	1	0.75	0.5	0.25
2	++++	1	1	0.75	0.5	0.25
3	+++	2	1	0.75	0.5	0.25
4	++	3	1	0.75	0.5	0.25
5	+	4	1	0.75	0.5	0.25

Table 10. Proposed negative weightage allocation for varying APTI values and under different environmental conditions.

Thus, the incorporation of the negative weightage in the calculation of the API for existing plant species can make it more practical if applied for the city like Delhi.

6. Conclusions

It can be concluded that due to the increasing pollution load on the plants, the vegetation in the city is under extreme stress. To ensure that the generated pollutant load is removed from the atmosphere, the quantification of the effect of the pollutants on the tree species should be made at regular intervals to ensure that they perform well under pollutant stresses. Further, the source apportionment studies have to be conducted to identify the pollutants like benzene which are exceeding the permissible standards during winter. Aditionally, the effect of the water contamination and other miscellaneous factors may add additional stress to the vegetation which needs to be considered to ensure that the species sustain in a polluted environment.

7. References

Bamnia, B.R., Kapoor, C.S., Kapoor, K and kapasya, V (2011) "Harmful effects of air pollution on physiological activites of Pongamia pinnita (L.) Pierre", Clean technology environmental policy, Springer. DOI 10.1007/s10098-011-0383-z.

CPCB (2007) "Phytoremediation of particulate matter from ambient environment through dust capturing plant species", Abstracted from www.cpcb.nic.in *on January 2nd*, *2012.*

CPCB (2010) "Air quality monitoring, emission inventory and source apportionment study for Indian cities", National Summary Report, Abstracted from www.moef.nic.in on January 5th , 2011.

DPCC (2011) "A Background note on status of PM_{10}, $PM_{2.5}$, NO_x and Benzene", Unpublished Report.

Govindaraju, M., Ganeshkumar, R.S., Muthukumaran, V.R and Visvanathan, P (2011) "Identification and evaluation of air-pollution-tolerant plants around lignite-based thermal power station for greenbelt development", Environmental Science Pollution Research,DOI 10.1007/s11356-011-0637-7.

Guenther, A., Hewitt, C.N., Erickson, D., Fall, R., Geron, C., Graedel, T., Harley, P., Klinger, L., Lerdan, M., Mckay, W.A., Pierce, T., Scholes, B., Steinbrecher, R., Tallamraju, R., Taylor, J. and Zimmerman, P (1995) "A global model of natural volatile organic emissions", Journal of Geophysical Research, 100, 8873–8892.

Khere, N., Mehta, V and Sabata, B.C (2009) "Interrelationships of birds and habitat features in urban green spaces in Delhi, India", Urban Foresty and Urban greeining, Volume 8, Issue 3, 2009, 187-196.

Lakshmi, P.S., K.L. Sravanti and N. Srinivas, (2009) Air pollution tolerance index of various plant species growing in industrial areas. The Ecoscan., 2: 203-206.

Lamb, B., Guenther, A., Gay, D., Westberg, H., (1987) "A nationalinventory of biogenic hydrocarbon emissions". Atmospheric Environment, 21, 1695–1705.

Li, M.H. (2003) "Peroxidase and superoxide dismutase activities in fig leaves in response to ambient air pollution in a subtropical city" Arch. Environ. Contamination toxicology, 45: 168-176.

Padhy and Varsheney (2005) "Isoprene emission from tropical tree species", Environ Pollut. 2005 May;135(1):101-9.

Rawat, J.S., Banerjee, S.P., (1996) "Urban forestry for improvement of environment' Journal of Energy Environment Monitoring 12 (2),109–116.

S.M. Seyyednjad, K. Majdian, H. Koochak and M. Niknejad, (2011) "Air Pollution Tolerance Indices of Some Plants Around Industrial Zone in South of Iran". Asian Journal of Biological Sciences, 4: 300-305.

Singh, S. K. and Rao, D. N. (1983). Evaluation of plants for their tolerance to air pollution. In Proceedings of the Symposium on Air Pollution Control, November, pp. 218-224.

Tewari, D.N., (1974) Urban forestry. Indian Forester 120 (8), 647–657.

Tingey, D.T., Manning, M., Grothaus, L.C., Burns, W.F., (1979) The influence of light and temperature on isoprene emission rates from live oak. Plant Physiology 47, 112–118.

3

Comprehensive Comparison of Trace Metal Concentrations in Inhaled Air Samples

Mehmet Yaman
Firat University, Science Faculty,
Department of Chemistry, Elazig
Turkey

1. Introduction

Because metals cannot be degraded or destroyed, the assessment of the health risks of metals via ambient air and dietary intake is an issue of special interest. Trace metals in air phase can be classified as metals or metalloids including the semi-metallic elements: boron, arsenic, selenium, and tellurium. Both natural and anthropogenic processes and sources emit metals and their compounds into the air. Anthropogenics; the processing of minerals, incineration of metallic objects, motor vehicle combustion of fuel containing metal additives, and the wearing out of motor vehicle tyres and brake pads result in the emission of metals associated with particulate matter. Metals occur naturally in soil and in rocks rich with minerals; thus weathering of the rocks, mining activities or even wind-blown dust can release these metals into air as particulate matter. Trace metals are part of a large group of air pollutants called air toxics, which upon inhalation or ingestion can be responsible for a range of health effects such as cancer, neurotoxicity, immunotoxicity, cardiotoxicity, reproductive toxicity, teratogenesis and genotoxicity (1-4).

When inhaled, very small particles containing metals or their compounds deposit beyond the bronchial regions of the lungs into the alveoli region. Epidemiological studies have established relationships between inhaled suspended particulate matter and morbidity/ mortality in populations (6-7). Studies in occupational or community settings have established the health effects of exposure to trace metals, such as lead, cadmium, nickel and their compounds (8-9). The accumulation of metals in human body can have middle and long-term health risks and can adversely affect the physiological functions (1-4). Metals can enter the human body mainly through inhalation and ingestion, with the diet being the main route of human exposure for non-occupationally exposed individuals. To evaluate and reduce the health and environmental effects of toxic metals in inhaled ambient air and food matrices, it is vitally important to know their chemical compositions and the way they vary in time and in space. Therefore, there are continuing efforts to determine particularly toxic metals such as Pb, Cd and Ni in air phases and food samples (10-14). In considering lead and cadmium in ambient air samples, this importance increases because the absorption rates of those metals by inhalation are significantly higher (up to 50-60%) than those by ingestion (between 3% and 10%) (10). The localized release of some heavy metals from inhaled particulate matter has been hypothesized to be responsible for the lung tissue damage.

In spite of all these facts, there are fewer studies available on Pb and Cd determinations in air samples compared to other food matrices due to, probably, the excessively lower concentrations of those metals in aerial matrix than the sensitivities of analysis methods (7, 9-12). In order to overcome those difficulties except using analytical techniques with high sensitivity such as electrothermal atomic absorption spectrometry (ETAAS), and inductively coupled plasma-mass spectrometry (ICP-MS), there are increased attentions to the usage of biomonitoring plants and plant parts such as leaves and shoots as biomonitoring (15-18).

2. Legislation

The emissions of three heavy metals, lead, mercury and cadmium, are being regulated in Europe under the Convention on Long-range Transboundary Air Pollution (19). This convention is the first international, legally binding instrument to deal with problems of air pollution on a broad regional basis. It covers 42 countries in Europe and North America and the European Union. Since entering into force in 1983, the Convention has been extended by several protocols dealing with specific pollutants. The Aarhus Protocol (20), the Protocol on Heavy Metals in June 1998, in Aarhus, Denmark (UN/ECE, 2000), targets three particularly harmful metals: cadmium, lead and mercury, to set a framework for national legislation that will lead to the substantial decrease in the emissions of the three metals in Europe and North America. The protocol seeks to cut emissions of heavy metals from industrial sources, combustion processes and waste incineration. The Protocol will enter into force when ratified by sixteen signatory countries; as of 3 July 2000, only six countries had ratified the Protocol (21).

Air toxics are not regulated under the National Environment Protection Measure (NEPM) for Ambient Air Quality, which addresses criteria pollutants in ambient air (22). However, a program initiated by the Commonwealth Government, the Living Cities-Air Toxics Program (ATP) was aimed at addressing urban air quality issues by supporting the development of national approaches to the management of 'air toxics'. For the purpose of the Living Cities initiative, air toxics are defined as: "...gaseous, aerosol or particulate *pollutants (other than the six criteria pollutants) which are present in the air in low concentrations with characteristics such as toxicity or persistence so as to be a hazard to human, plant or animal life...*" (23). The terms 'air toxics' and 'hazardous air pollutants' (HAPs) are used interchangeably (24). The Technical Advisory Group (TAG) for the ATP included the metals cadmium, chromium (VI), mercury, nickel and their compounds in the list of 28 priority air toxics identified in the ATP. Further, lead is the most routinely monitored heavy metal in ambient air in some countries as a result of its presence in motor vehicle fuel (25).

3. Speciation and toxicity

Compared to gaseous compounds, the assessment of metal and metalloid compounds in inhaled ambient air is complicated by the fact that different species with considerably differing toxicity and/or carcinogenic potency may be encountered. In order to fully evaluate the health effects, it is important to know which species do occur in the environment or at least which compounds form the main constituents. In ambient air, metals, metalloids and their compounds are mainly encountered as part of particulate matter. They may be present in the non soluble, non stoichiometric mixture phase such as spinels or as soluble ionic compounds such as salts. In respect to their effects on the environment and on human health, gaseous forms such as organometallic compounds can be characterized by other parameters, such as water solubility (extended to solubility in biological fluids), particle size distribution, morphology and specific surface area, and chemical heterogeneity of the particles, or the concentration of metals and metalloids in the particles ultimately contacting target tissues in the human body (26). For example, a metal compound is encapsulated in another aerosol or surface enrichment of volatile species.

All parameters mentioned will influence the bioavailability and possible effects. In addition, metal and metalloid containing substances can undergo various chemical and physical transformations in the atmosphere on their way from the source to a possible receptor. For example, As (III) compounds may be oxidized to As (V). Unfortunately, analytical methods normally only identify the elements which are present in atmospheric particles, species specific analysis being extremely difficult in the concentration range occurring in ambient air (typically several ng/m³. In addition, the state of oxidation may change during sampling. Consequently, information on the concentration of different species in ambient air is very limited at present. Further, the limited knowledge available on the occurrence of species is outlined.

Trace elements are found naturally in the environment and human exposure derives from a variety of sources, including air, drinking water, and food. Concentrations of trace elements in the air are generally low. Levels of As in the air range from approximately 1 to 2,000 ng/m³, levels of Cd generally range from 1 to 40 ng/m³ but can reach up to 100 ng/m³ near emission sources, and levels of Ni in cities and rural areas range from 7 to 12 ng/m³ (27-54). Workers in the smelting and refining industries and those employed in the production of batteries, coatings, and plastics can be exposed to much higher levels of airborne Cd and Ni (55).

3.1 Trace metal uses and their health effects

3.1.1 Arsenic

There are three major groups of arsenic (As) compounds: inorganic arsenic compounds, organic arsenic compounds, arsine gas and substituted arsines.

Elemental arsenic is utilized in alloys in order to increase their hardness and heat resistance. It is also used in the manufacture of certain types of glass, as a component of electrical devices and as a doping agent in germanium and silicon solid-state products. The uses and source of arsenic compounds are summarized in Table 1.

Organic arsenic compounds in marine organisms occur in concentrations corresponding to a concentration of arsenic in the range 1 to 100 mg/kg in marine organisms such as shrimp and fish. Such arsenic is mainly made up of arsenobetaine and arsenocholine, organic arsenic compounds of low toxicity (56). The substituted arsines are trivalent organic arsenical compounds which, depending on the number of alkyl or phenyl groups that they have attached to the arsenic nucleus, are known as mono-, di- or tri-substituted arsines. Dichloroethylarsine ($C_2H_5AsCl_2$), or ethyldichloroarsine, is a colourless liquid with an irritant odour. This compound was developed as a potential chemical warfare agent. Dichloro(2-chlorovinyl-) arsine ($ClCH:CHAsCl_2$), or chlorovinyldichloroarsine (lewisite), is an olive-green liquid with a germanium-like odour. It was developed as a potential warfare agent but never used according to our knowledge. Dimethyl-arsine $(CH_3)_2AsH$, or cacodyl hydride and trimethylarsine $(CH_3)_3As$), or trimethylarsenic, are both colourless liquids. These two compounds can be produced after metabolic transformation of arsenic compounds by bacteria and fungi.

Compound	Uses/Source
Arsenic trichloride $(AsCl_3)$	ceramics industry, manufacturing of chlorine-containing arsenicals
Arsenic trioxide (As_2O_3), or white arsenic	purification of synthesis gas, as a primary material for all arsenic compounds, preservative for hides and wood, a textile mordant, a reagent in mineral flotation, a decolourizing and refining agent in glass manufacture
Calcium arsenite $(Ca(As_2H_2O_4)$, Calcium arsenate $(Ca_3(AsO_4)_2)$	insecticides
Cacodylic acid $((CH_3)_2AsOOH)$	herbicide and a defoliant
cupric acetoarsenite (usually considered $Cu(COOCH_3)_2\ 3Cu(AsO_2)_2)$	Insecticides, for painting ships and submarines
Sodium arsenite $(NaAsO_2)$	herbicide, a corrosion inhibitor, as a drying agent in the textile industry
Arsenic trisulphide	a component of infrared-transmitting glass, a dehairing agent in the tanning industry, the manufacturing of pyrotechnics and semiconductors
Arsenic acid $(H_3AsO_4\ \cdot\tfrac{1}{2}H_2O)$	the manufacturing of arsenates, glass making, wood-treating processes
Arsenic pentoxide (As_2O_5)	Herbicide, a wood preservative, in the manufacture of coloured glass
Arsanilic acid $(NH_2C_6H_4AsO(OH)_2)$	as a grasshopper bait, as an additive in animal feeds
Arsine gas	in organic syntheses, in the processing of solid-state electronic components, inadvertently in industrial processes when nascent hydrogen is formed and arsenic is present

Table 1. Arsenic compounds and their uses/source.

3.1.1.1 Toxicity

Although it is possible that very small amounts of certain arsenic compounds may have beneficial effects, as indicated by some animal studies, arsenic compounds, particularly the inorganic ones, are otherwise regarded as very potent poisons. Acute toxicity varies widely among compounds, depending on their valency state and solubility in biological media. The soluble trivalent compounds are the most toxic. Uptake of inorganic arsenic compounds from the gastrointestinal tract is almost complete, but uptake may be delayed for less soluble forms such as arsenic trioxide in particle form. Uptake after inhalation is also almost complete, since even less soluble material deposited on the respiratory mucosa, will be transferred to the gastrointestinal tract and subsequently taken up. The health effects of arsenic were summarized in Table 2.

Metal-Route of exposure	Health effects	Diagnosis/medical monitoring
Inorganic and organic As-inhalation, ingestion, skin	- Acute exposure: nausea, diarrhea, Gastro intestinal (GI) bleeding, cardiovascular effects, shock, and death. Liver, kidney damage and seizures have been reported. - Chronic exposure: hyperpigmentation of skin, warts, corns, heart disease, neuropathy, liver damage, peripheral vascular disease (gangrene of lower limbs), and increased risk of skin, liver, lung and bladder cancer. Arsenic in drinking water can also cause diabetes and hypertension. - Anaemia, leucopenia and granulocytopenia - The soluble trivalent compounds are the most toxic - The fatal dose of ingested arsenic trioxide has been reported to range from 70 to 180 mg. Death may occur within 24 hours, but the usual course runs from 3 to 7 days. - Cancer of the respiratory tract has been reported in excess frequency among workers engaged in the production of insecticides containing lead arsenate and calcium arsenate, in vine-growers spraying insecticides containing inorganic copper and arsenic compounds, and in smelter workers exposed to inorganic compounds of arsenic and a number of other metals. The latency time between onset of exposure and the appearance of cancer is long, usually between 15 and 30 years. - Arsine is one of the most powerful haemolytic agents found in industry. Inhalation of 250 ppm of arsine gas is instantly lethal. Exposure from 25 to 50 ppm for 30 minutes is lethal, and 10 ppm may be lethal after longer exposures.	Urinary arsenic level is the most reliable indicator of recent exposure to arsenic. Arsenic in hair and fingernails can indicate exposure to high levels in the past 6–12 months.
Cd-inhalation, ingestion	- Cell proliferation, differentiation, apoptosis, and other cellular activities, numerous molecular lesions caused carcinogenesis. - Cadmium targets the lung, liver, kidney, and testes. - In acute intoxication: nephrotoxicity, immunotoxicity, osteotoxicity, and tumors after prolonged exposures. - Prostate, lung, testicular, renal, and skeletal cancers. - In vivo: Generate $O_2^{\bullet-}$, H_2O_2, and $\bullet OH$ accompanied by activation of redox-sensitive transcription factors. - After inhalation above 1 mg Cd/m^3 in air for 8 hours; chemical pneumonitis, and in severe cases pulmonary oedema.	Cadmium levels in blood are mainly an indication of the last few months exposure, but can be used to assess body burden a few years after exposure has ceased. The individual critical concentrations of

Metal-Route of exposure	Health effects	Diagnosis/medical monitoring
Cd-inhalation, ingestion	- After ingestion of drinks exceeding 15 mg Cd/l; nausea, vomiting, abdominal pains and sometimes diarrhoea. - Prolonged exposure in air at concentrations exceeding 0.1 mg Cd/m^3; Pulmonary emphysema - Exposure for more than 20 years to concentrations of about 0.02 mg Cd/m^3, certain pulmonary effects. - Exceeding 200 μg Cd/g (this is critical concentration) wet weight of renal cortex; tubular dysfunction with decreased reabsorption of proteins from the urine, tubular proteinuria with increased excretion of low-molecular-weight proteins. - The average cadmium concentration in workroom air (8 hours per day) should not exceed 0.01 mg Cd/m^3.	cadmium in urine and/or in blood are 50 nmol/l whole blood or 3 nmol/mmol creatinine.
Lead-Inhalation, ingestion, skin	- Hematologic: decreased heme synthesis enzymes, anemia. - Cardiovascular: elevated blood pressure. - Cognitive, neurobehavioral, and psychological effects. - Gastrointestinal: colic or abdominal cramps. - Peripheral neuropathy; encephalopathy (at high levels). - Reduced fertility. - Immune system: alterations in T cell, reduced IgG serum levels. - Children: lethargy, loss of appetite, anemia, colic, neurological impairment, and impaired metabolism of Vit D. - Exposure in uterus and during childhood can result in impaired neurological development, IQ deficits, and growth retardation.	Lead in whole blood is a reliable test. Erythrocyte protoporphyrin (EP) test can also be used but it is not sensitive to detect high levels of lead in children.
Mercury-Inhalation, ingestion	- All forms of mercury are toxic to the central nervous system (CNS). - Exposure to high levels can damage brain, kidneys, and developing fetus. (methyl mercury is the most toxic form). - Toxicity to brain results in irritability, tremors, visual changes, and memory problems. - Mercury salts can cause abdominal cramps, diarrhea, and kidney damage.	Acute exposure is best measured by mercury in blood and chronic exposure by mercury in urine

Metal- Route of exposure	Health effects	Diagnosis/medical monitoring
Nickel- Inhalation, ingestion	- The National Maximum Workplace Concentration Committee (NMWCC) of the Netherlands proposed that urine nickel concentration 40 μg/g creatinine, or serum nickel concentration 5 μg/l (both measured in samples obtained at the end of a working week or a work shift) be considered warning limits for further investigation of workers exposed to nickel metal or soluble nickel compounds. - Exposures are classified as "mild" if the initial 8-h specimen of urine has a nickel concentration less than 100 μg/l, "moderate" if the nickel concentration is 100 to 500 μg/l, and "severe" if the nickel concentration exceeds 500 μg/l. - Chronic exposure of workers to inhalation of low atmospheric concentrations of nickel carbonyl (0.007 to 0.52 mg/m^3) can cause neurological symptoms such as insomnia, headache, dizziness, memory loss, and other manifestations including chest tightness, excessive sweating, alopecia.	

Table 2. Reported metal toxicity and diagnosis/medical monitoring.

Occupational exposure to inorganic arsenic compounds through inhalation, ingestion or skin contact with subsequent absorption may occur in industry. Acute effects at the point of entry may occur if exposure is excessive. Dermatitis may occur as an acute symptom but is more often the result of toxicity from long-term exposure, sometimes subsequent to sensitization.

In occupational exposure to mainly airborne arsenic, skin lesions may result from local irritation. Two types of dermatological disorders may occur:

1. an eczematous type with erythema (redness), swelling and papules or vesicles
2. a follicular type with erythema and follicular swelling or follicular pustules.

Dermatitis is primarily localized on the most heavily exposed areas, such as the face, back of the neck, forearms, wrists and hands. Patch tests have demonstrated that the dermatitis is due to arsenic, not to impurities present in the crude arsenic trioxide. Chronic dermal lesions may follow this type of initial reaction, depending on the concentration and duration of exposure. These chronic lesions may occur after many years of occupational or environmental exposure. Hyperkeratosis, warts and melanosis of the skin are the conspicuous signs (57).

3.1.1.2 Carcinogenic effects

Inorganic arsenic compounds are classified by the International Agency for Research on Cancer (IARC) as lung and skin carcinogens (58). There is also some evidence to suggest that persons exposed to inorganic arsenic compounds suffer a higher incidence of angiosarcoma of the liver and possibly of stomach cancer. A synergistic action of tobacco smoking has been demonstrated for lung cancer. Long-term exposure to inorganic arsenic via drinking water has been associated with an increased incidence of skin cancer. This increase has been shown to be related to concentration in drinking water.

3.1.1.3 Organic arsenic compounds

Organic arsenicals used as pesticides or as drugs may also give rise to toxicity, although such adverse effects are incompletely documented in humans. Toxic effects on the nervous system have been reported in experimental animals following feeding with high doses of arsanilic acid, which is commonly used as a feed additive in poultry and swine.

The organic arsenic compounds that occur in foodstuffs of marine origin, such as shrimp, crab and fish, are made up of arsinocholine and arsinobetaine. It is well known that the amounts of organic arsenic that are present in fish and shellfish can be consumed without ill effects because these compounds are quickly excreted, mainly via urine.

Many cases of acute arsine poisoning have been recorded, and there is a high fatality rate. Arsine is one of the most powerful haemolytic agents found in industry. Its haemolytic activity is due to its ability to cause a fall in erythrocyte-reduced glutathione content. Signs and symptoms of arsine poisoning include haemolysis, which develops after a latent period that is dependent on the intensity of exposure. Inhalation of 250 ppm of arsine gas is instantly lethal. Exposure from 25 to 50 ppm for 30 minutes is lethal, and 10 ppm may be lethal after longer exposures. The signs and symptoms of poisoning are those characteristic of an acute and massive haemolysis. After acute and severe exposure, a peripheral neuropathy may develop and can still be present several months after poisoning. Little is known about repeated or chronic exposure to arsine, but since the arsine gas is metabolized to inorganic arsenic in the body, it can be assumed that there is a risk for symptoms similar to those in long-term exposure to inorganic arsenic compounds (59).

3.1.2 Cadmium

Cadmium (Cd) has many chemical and physical similarities to zinc and occurs together with zinc in nature. In minerals and ores, cadmium and zinc generally have a ratio of 1:100 to 1:1,000. Cadmium is highly resistant to corrosion and has been widely used for electroplating of other metals, mainly steel and iron. Screws, screw nuts, locks and various parts for aircraft and motor vehicles are frequently treated with cadmium in order to withstand corrosion. Nowadays, however, only 8% of all refined cadmium is used for platings and coatings. It was established that Cd of used in developed and industrialized countries is about 3% in certain alloys, 8% for platings and coatings, 30% for pigments and stabilizers in plastics, 55% for rechargeable, small portable cadmium-containing batteries used in mobile telephones and similars.

The most important Cd compound is cadmium stearate, which is used as a heat stabilizer in polyvinyl chloride (PVC) plastics. Cadmium sulphide and cadmium sulphoselenide are used as yellow and red pigments in plastics and colours. Cadmium sulphide is also used in photo- and solar cells. Cadmium chloride acts as a fungicide, an ingredient in elecroplating baths, a colourant for pyrotechnics, an additive to tinning solution and a mordant in dyeing and printing textiles. It is also used in the production of certain photographic films and in the manufacture of special mirrors and coatings for electronic vacuum tubes. Cadmium oxide is an elecroplating agent, a starting material for PVC heat stabilizers and a component of silver alloys, phosphors, semiconductors and glass and ceramic glazes (60).

As a result, cadmium can represent an environmental hazard, and many countries have introduced legislative actions aimed towards decreasing the use and subsequent environmental spread of cadmium.

3.1.2.1 Toxicity

The health effects of cadmium were summarized in Table 2. Metallothioneins play a role in the homeostasis of essential metals such as copper, detoxification of toxic metals such as cadmium, and protection against oxidative stress. Gastrointestinal absorption of ingested cadmium is about 2 to 6% under normal conditions. Individuals with low body iron stores, reflected by low concentrations of serum ferritin, may have considerably higher absorption of cadmium, up to 20% of a given dose of cadmium. Significant amounts of cadmium may also be absorbed via the lung from the inhalation of tobacco smoke or from occupational exposure to atmospheric cadmium dust. Pulmonary absorption of inhaled respirable cadmium dust is estimated at 20 to 50%. After absorption via the gastrointestinal tract or the lung, cadmium is transported to the liver, where production of a cadmium-binding low-molecular-weight protein, metallothionein, is initiated (61).

About 80 to 90% of the total amount of cadmium in the body is considered to be bound to metallothionein. This prevents the free cadmium ions from exerting their toxic effects. It is likely that small amounts of metallothionein-bound cadmium are constantly leaving the liver and being transported to the kidney via the blood. The metallothionein with the cadmium bound to it is filtered through the glomeruli into the primary urine. Like other low-molecular-weight proteins and amino acids, the metallothionein-cadmium complex is subsequently reabsorbed from the primary urine into the proximal tubular cells, where digestive enzymes degrade the engulfed proteins into smaller peptides and amino acids. Free cadmium ions in the cells result from degradation of metallothionein and initiate a new synthesis of metallothionein, binding the cadmium, and thus protecting the cell from the highly toxic free cadmium ions. Kidney dysfunction is considered to occur when the metallothionein-producing capacity of the tubular cells is exceeded. The kidney and liver have the highest concentrations of cadmium, together containing about 50% of the body burden of cadmium. The cadmium concentration in the kidney cortex, before cadmium-induced kidney damage occurs, is generally about 15 times the concentration in liver. Elimination of cadmium is very slow. As a result of this, cadmium accumulates in the body, the concentrations increasing with age and length of exposure (62). Based on organ concentration at different ages the biological half-life of cadmium in humans has been estimated in the range of 7 to 30 years.

3.1.2.2 Acute toxicity

Inhalation of cadmium compounds at concentrations above 1 mg Cd/m^3 in air for 8 hours, or at higher concentrations for shorter periods, may lead to chemical pneumonitis, and in severe cases pulmonary oedema. Symptoms generally occur within 1 to 8 hours after exposure. They are influenza-like and similar to those in metal fume fever. The more severe symptoms of chemical pneumonitis and pulmonary oedema may have a latency period up to 24 hours. Death may occur after 4 to 7 days. Exposure to cadmium in the air at concentrations exceeding 5 mg Cd/m3 is most likely to occur where cadmium alloys are smelted, welded or soldered. Ingestion of drinks contaminated with cadmium at concentrations exceeding 15 mg Cd/l gives rise to symptoms of food poisoning. Symptoms are nausea, vomiting, abdominal pains and sometimes diarrhoea. Sources of food contamination may be pots and pans with cadmium-containing glazing and cadmium solderings used in vending machines for hot and cold drinks. In animals parenteral administration of cadmium at doses exceeding 2 mg Cd/kg body weight causes necrosis of the testis. No such effect has been reported in humans.

3.1.2.3 Chronic toxicity

Chronic cadmium poisoning has been reported after prolonged occupational exposure to cadmium oxide fumes, cadmium oxide dust and cadmium stearates. Changes associated with chronic cadmium poisoning may be local, in which case they involve the respiratory tract, or they may be systemic, resulting from absorption of cadmium. Systemic changes include kidney damage with proteinuria and anemia. Lung disease in the form of emphysema is the main symptom at heavy exposure to cadmium in air, whereas kidney dysfunction and damage are the most prominent findings after long-term exposure to lower levels of cadmium in workroom air or via cadmium-contaminated food. Mild hypochromic anemia is frequently found among workers exposed to high levels of cadmium. This may be due to both increased destruction of red blood cells and to iron deficiency. Yellow discolouration of the necks of teeth and loss of sense of smell (anosmia) may also be seen in cases of exposure to very high cadmium concentrations.

Pulmonary emphysema is considered a possible effect of prolonged exposure to cadmium in air at concentrations exceeding 0.1 mg Cd/m³. It has been reported that exposure to concentrations of about 0.02 mg Cd/m³ for more than 20 years can cause certain pulmonary effects. Cadmium-induced pulmonary emphysema can reduce working capacity and may be the cause of invalidity and life shortening. With long-term low-level cadmium exposure the kidney is the critical organ (i.e., the organ first affected). Cadmium accumulates in renal cortex. Concentrations exceeding 200 μg Cd/g wet weight have previously been estimated to cause tubular dysfunction with decreased reabsorption of proteins from the urine (63). This causes tubular proteinuria with increased excretion of low-molecular-weight proteins such as α,α-1-microglobulin (protein HC), β-2-microglobulin and retinol binding protein (RTB). Recent research suggests, however, that tubular damage may occur at lower levels of cadmium in kidney cortex. As the kidney dysfunction progresses, amino acids, glucose and minerals, such as calcium and phosphorus, are also lost into the urine. Increased excretion of calcium and phosphorous may disturb bone metabolism, and kidney stones are frequently reported by cadmium workers. After long-term medium-to-high levels of exposure to cadmium, the kidney's glomeruli may also be affected, leading to a decreased glomerular filtration rate. In severe cases uraemia may develop. Excessive cadmium exposure has occurred in the general population through ingestion of contaminated rice and other foodstuffs, and possibly drinking water. The itai-itai disease, a painful type of osteomalacia, with multiple fractures appearing together with kidney dysfunction, has occurred in Japan in areas with high cadmium exposure. Though the pathogenesis of itai-itai disease is still under dispute, it is generally accepted that cadmium is a necessary aetiological factor. It should be stressed that cadmium-induced kidney damage is irreversible and may grow worse even after exposure has ceased.

3.1.2.4 Carcinogenic effects

Cd competes with Zn for binding sites and can therefore interfere with some of Zinc's essential functions. Thus, it may inhibit enzyme reactions and utilization of nutrients. Cd can generate free radical tissue damage because it may be a catalyst to oxidation reactions. Furthermore, excessive Cd exposure can cause renal damage, reproduction problems, cardiovascular diseases and hypertension. There are several sources of human exposure to Cd, including employment in primary metal industries, production of certain batteries, some electroplating processes and consumption of tobacco products. Consequently, it was reported by International Agency Research on Cancer (IARC) that through inhalation cadmium could cause lung cancer in humans and animals (64). As a result, the World Health Organization (WHO) (65) established provisional tolerable weekly intakes (PTWIs) of Cd of 0.007 microgram/kg body weight, for all human groups.

There is strong evidence of dose-response relationships and an increased mortality from lung cancer in several epidemiological studies on cadmium-exposed workers. The interpretation is complicated by concurrent exposures to other metals which are known or suspected carcinogens. Continuing observations of cadmium-exposed workers have, however, failed to yield evidence of increased mortality from prostatic cancer, as initially suspected. The IARC in 1993 (64) assessed the risk of cancer from exposure to cadmium and concluded that it should be regarded as a human carcinogen. Since then additional epidemiological evidence has come forth with somewhat contradictory results, and the possible carcinogenicity of cadmium thus remains unclear. It is nevertheless clear that cadmium possesses strong carcinogenic properties in animal experiments.

3.1.2.5 Limitations

The kidney cortex is the critical organ with long-term cadmium exposure via air or food. The critical concentration is estimated at about 200 µg Cd/g wet weight, but may be lower, as stated above. In order to keep the kidney cortex concentration below this level even after lifelong exposure, the average cadmium concentration in workroom air (8 hours per day) should not exceed 0.01 mg Cd/m^3.

To ensure that excessive accumulation of cadmium in the kidney does not occur, cadmium levels in blood and in urine should be checked regularly. Cadmium levels in blood are mainly an indication of the last few months exposure, but can be used to assess body burden a few years after exposure has ceased. A value of 100 nmol Cd/l whole blood is an approximate critical level if exposure is regular for long periods. Cadmium values in urine can be used to estimate the cadmium body burden, providing kidney damage has not occurred. It has been estimated by the WHO that 10 nmol/mmol creatinine is the concentration below which kidney dysfunction should not occur. Recent research has, however, shown that kidney dysfunction may occur already at around 5 nmol/mmol creatinine. Since the mentioned blood and urinary levels are at levels at which action of cadmium on kidney has been observed, it is recommended that control measures should be applied whenever the individual concentrations of cadmium in urine and/or in blood exceed 50 nmol/l whole blood or 3 nmol/mmol creatinine respectively.

3.1.3 Chromium

In addition to chromic acid, the ferrous chromite ($FeOCr_2O_3$) ore contains variable quantities of other substances. Only ores or concentrates containing more than 40% chromic oxide (Cr_2O_3) are used commercially, and countries having the most suitable deposits are the Russian Federation, South Africa, Zimbabwe, Turkey, the Philippines and India. The prime consumers of chromites are the United States, the Russian Federation, Germany, Japan, France and the United Kingdom.

The most significant usage of pure chromium is for electroplating of a wide range of equipment, such as automobile parts and electric equipment. Chromium is used extensively for alloying with iron and nickel to form stainless steel, and with nickel, titanium, niobium, cobalt, copper and other metals to form special-purpose alloys.

Chromium forms a number of compounds in various oxidation states. Those of II (chromous), III (chromic) and VI (chromate) states are most important; the II state is basic, the III state is amphoteric and the VI state is acidic. Commercial applications mainly concern compounds in the VI state, with some interest in III state chromium compounds.

The chromous state (Cr^{II}) is unstable and is readily oxidized to the chromic state (Cr^{III}). This instability limits the use of chromous compounds. The most important compounds containing chromium in the Cr^{VI} state are dichromate compounds and chromium trioxide.

Compounds containing Cr^{VI} are used in many industrial operations: the manufacture of important inorganic pigments such as lead chromes, molybdate-oranges, zinc chromate and chromium-oxide green; wood preservation; corrosion inhibition; and coloured glasses and glazes. Basic chromic sulphates are widely used for tanning. The dyeing of textiles, the preparation of many important catalysts containing chromic oxide and the production of light-sensitive dichromated colloids for use in lithography are also well-known industrial uses of chromium-containing chemicals.

Chromic acid is used not only for "decorative" chromium plating but also for "hard" chromium plating, where it is deposited in much thicker layers to give an extremely hard surface with a low coefficient of friction.

Because of the strong oxidizing action of chromates in acid solution, there are many industrial applications particularly involving organic materials, such as the oxidation of trinitrotoluene (TNT) to give phloroglucinol and the oxidation of picoline to give nicotine acid (66).

3.1.3.1 Toxicity

Compounds with Cr^{III} oxidation states are considerably less hazardous than are Cr^{VI} compounds. Compounds of Cr^{III} are poorly absorbed from the digestive system. These Cr^{III} compounds may also combine with proteins in the superficial layers of the skin to form stable complexes. Compounds of Cr^{III} do not cause chrome ulcerations and do not generally initiate allergic dermatitis without prior sensitization by Cr^{VI} compounds.

In the Cr^{VI} oxidation state, chromium compounds are readily absorbed after ingestion as well as during inhalation. The uptake through intact skin is less well elucidated. The irritant and corrosive effects caused by Cr^{VI} occur readily after uptake through mucous membranes, where they are readily absorbed. Work-related exposure to Cr^{VI} compounds may induce skin and mucous membrane irritation or corrosion, allergic skin reactions or skin ulcerations.

The untoward effects of chromium compounds generally occur among workers in workplaces where Cr^{VI} is encountered, in particular during manufacture or use (67). The effects frequently involve the skin or respiratory system. Typical industrial hazards are inhalation of the dust or fumes arising during the manufacture of dichromate from chromite ore and the manufacture of lead and zinc chromates, inhalation of chromic acid mists during electroplating or surface treatment of metals, and skin contact with Cr^{VI} compounds in manufacture or use. Exposure to Cr^{VI}-containing fumes may also occur during welding of stainless steels.

Numerous sources of exposure to Cr^{VI} can be listed as contact with cement, plaster, leather, graphic work, work in match factories, work in tanneries and various sources of metal work. Workers employed in wet sandpapering of car bodies have also been reported with allergy. Affected subjects react positively to patch testing with 0.5% dichromate.

It has been shown that Cr^{VI} penetrates the skin through the sweat glands and is reduced to Cr^{III} in the corium. It is shown that the Cr^{III} then reacts with protein to form the antigen-antibody complex. This explains the localization of lesions around sweat glands and why very small amounts of dichromate can cause sensitization. The chronic character of the dermatitis may be due to the fact that the antigen-antibody complex is removed more slowly than would be the case if the reaction occurred in the epidermis.

Inhalation of dust or mist containing Cr^{VI} is irritating to mucous membranes. At high concentrations of such dust, sneezing, rhinorrhoea, lesions of the nasal septum and redness of the throat are documented effects. Sensitization has also been reported, resulting in typical asthmatic attacks, which may recur on subsequent exposure. At exposure for several days to chromic acid mist at concentrations of about 20 to 30 mg/m^3, cough, headache, dyspnoea and substernal pain have also been reported after exposure. The occurrence of bronchospasm in a person working with chromates should suggest chemical irritation of the lungs.

In previous years, when the exposure levels to Cr^{VI} compounds could be high, ulcerations of the nasal septum were frequently seen among exposed workers. This untoward effect results from deposition of Cr^{VI}-containing particulates or mist droplets on the nasal septum, resulting in ulceration of the cartilaginous portion followed, in many cases, by perforation at the site of ulceration. Frequent nose-picking may enhance the formation of perforation.

Necrosis of the kidneys has also been reported, starting with tubular necrosis, leaving the glomeruli undamaged. Diffuse necrosis of the liver and subsequent loss of architecture has also been reported. Soon after the turn of the century there were a number of reports on human ingestion of Cr^{VI} compounds resulting in major gastro-intestinal bleeding from ulcerations of the intestinal mucosa. Sometimes such bleedings resulted in cardiovascular shock as a possible complication. If the patient survived, tubular necrosis of the kidneys or liver necrosis could occur.

Increased incidence of lung cancer among workers in manufacture and use of Cr^{VI} compounds has been reported in a great number of studies from France, Germany, Italy,

Japan, Norway, the United States and the United Kingdom (68). Chromates of zinc and calcium appear to be among the most potent carcinogenic chromates, as well as among the most potent human carcinogens. Elevated incidence of lung cancer has also been reported among subjects exposed to lead chromates, and to fumes of chromium trioxides. Heavy exposures to Cr^{VI} compounds have resulted in very high incidence of lung cancer in exposed workers 15 or more years after first exposure, as reported in both cohort studies and case reports.

Thus, it is well established that an increase in the incidence of lung cancer of workers employed in the manufacture of zinc chromate and the manufacture of mono- and dichromates from chromite ore is a long-term effect of work-related heavy exposure to Cr^{VI} compounds. Some of the cohort studies have reported measurements of exposure levels among the exposed cohorts. Also, a small number of studies have indicated that exposure to fumes generated from welding on Cr-alloyed steel may result in elevated incidence of lung cancer among these welders.

There is no firmly established "safe" level of exposure. However, most of the reports on association between Cr^{VI} exposure and cancer of the respiratory organs and exposure levels report on air levels exceeding 50 mg Cr^{VI}/m^3 air (69).

Water-soluble, acid soluble and water insoluble chromium is found in the lung tissues of chromate workers in varying amounts.

Although it has not been firmly established, some studies have indicated that exposure to chromates may result in increased risk of cancer in the nasal sinuses and the alimentary tract. The studies that indicate excess cancer of the alimentary tract are case reports from the 1930s or cohort studies that reflect exposure at high levels than generally encountered today.

3.1.4 Iron

In addition to ferroalloys, the most important industrial iron compounds are the oxides and the carbonate, which constitute the principal ores from which the metal is obtained. Of lesser industrial importance are cyanides, nitrides, nitrates, phosphides, phosphates and iron carbonyl.

3.1.4.1 Toxicity

Industrial dangers are present during the mining, transportation and preparation of the ores, during the production and use of the metal and alloys in iron and steel works and in foundries, and during the manufacture and use of certain compounds. Inhalation of iron dust or fumes occurs in iron-ore mining; arc welding; metal grinding, polishing and working; and in boiler scaling. If inhaled, iron is a local irritant to the lung and gastrointestinal tract. Reports indicate that long-term exposure to a mixture of iron and other metallic dusts may impair pulmonary function. Inhaling dust containing silica or iron oxide can lead to pneumoconiosis, but there are no definite conclusions as to the role of iron oxide particles in the development of lung cancer in humans. Based on animal experiments, it is suspected that iron oxide dust may serve as a "co-carcinogenic" substance, thus enhancing the development of cancer when combined simultaneously with exposure to carcinogenic substances (70).

Mortality studies of haematite (Fe_2O_3) miners (containing up to 66% iron) have shown an increased risk of lung cancer, generally among smokers, in several mining areas such as Cumberland, Lorraine, Kiruna and Krivoi Rog. In experimental studies, ferric oxide has not been found to be carcinogenic; however, the experiments were not carried out with haematite (71). The presence of radon in the atmosphere of haematite mines has been suggested to be an important carcinogenic factor. Epidemiological studies of iron and steel foundry workers have typically noted risks of lung cancer elevated by 1.5- to 2.5-fold. The International Agency for Research on Cancer (IARC) classifies iron and steel founding as a carcinogenic process for humans. The specific chemical agents involved (e.g., polynuclear aromatic hydrocarbons, silica, metal fumes) have not been identified. An increased incidence of lung cancer has also been reported, but less significantly, among metal grinders. The conclusions for lung cancer among welders are controversial. The dangerous properties of the remaining iron compounds are usually due to the radical with which the iron is associated. Thus ferric arsenate ($FeAsO_4$) and ferric arsenite ($FeAsO_3 \cdot Fe_2O_3$) possess the poisonous properties of arsenical compounds. Iron carbonyl ($FeCO_5$) is one of the more dangerous of the metal carbonyls, having both toxic and flammable properties (71).

Ferrosilicon production can result in both aerosols and dusts of ferrosilicon. Animal studies indicate that ferrosilicon dust can cause thickening of the alveolar walls with the occasional disappearance of the alveolar structure. The raw materials used in alloy production may also contain free silica, although in relatively low concentrations. There is some disagreement as to whether classical silicosis may be a potential hazard in ferrosilicon production. There is no doubt, however, that chronic pulmonary disease, whatever its classification, can result from excessive exposure to the dust or aerosols encountered in ferrosilicon plants.

3.1.5 Lead

About 40% of lead is used as a metal, 25% in alloys and 35% in chemical compounds. Because populations in, at least, 100 countries are still exposed to air pollution with lead in spite of banning the usage of lead in gasoline in many countries, usage of lead and lead compounds will be detailed. Due to its malleability, low melting point, and ability to form compounds, Pb has been used in hundreds of products such as pipes, solder, brass fixtures, crystal, paint, cable, ceramics, and batteries (72). Metallic lead is used in the form of sheeting or pipes where pliability and resistance to corrosion are required, such as in chemical plants and the building industry; it is used also for cable sheathing, as an ingredient in solder and as a filler in the automobile industry. It is a valuable shielding material for ionizing radiations. It is used for metallizing to provide protective coatings, in the manufacture of storage batteries and as a heat treatment bath in wire drawing. Lead is present in a variety of alloys and its compounds are prepared and used in large quantities in many industries. Lead oxides are used in the plates of electric batteries and accumulators (PbO and Pb_3O_4), as compounding agents in rubber manufacture (PbO), as paint ingredients (Pb_3O_4) and as constituents of glazes, enamels and glass.

Lead salts form the basis of many paints and pigments; lead carbonate and lead sulphate are used as white pigments and the lead chromates provide chrome yellow, chrome orange, chrome red and chrome green. Lead arsenate is an insecticide, lead sulphate is used in rubber compounding, lead acetate has important uses in the chemical industry, lead naphthenate is an extensively used dryer and tetraethyllead is an antiknock additive for gasoline, where still permitted by law.

Other metals such as antimony, arsenic, tin and bismuth may be added to lead to improve its mechanical or chemical properties, and lead itself may be added to alloys such as brass, bronze and steel to obtain certain desirable characteristics. The very large numbers of organic and inorganic lead compounds are encountered in industry.

3.1.5.1 Toxicity

The prime hazard of lead is its toxicity. For a long time, it is known that lead is toxic for brain, kidney and reproductive system and can also cause impairment in intellectual functioning, infertility, miscarriage and hypertension. Several studies have shown that lead exposures in school-aged children can significantly reduce IQ and has been associated with aggressive behavior, delinquency and attention disorders (73). The health effects of lead were summarized in Table 2.

Clinical lead poisoning has always been one of the most important occupational diseases. Industrial consumption of lead is increasing and traditional consumers are being supplemented by new users such as the plastics industry. Hazardous exposure to lead, therefore, occurs in many occupations. In lead mining, a considerable proportion of lead absorption occurs through the alimentary tract and consequently the extent of the hazard in this industry depends, to some extent, on the solubility of ores being worked. The lead sulphide (PbS) in galena is insoluble and absorption from the lung is limited; however, in the stomach, some lead sulphide may be converted to slightly soluble lead chloride which may then be absorbed in moderate quantities. In lead smelting, the main hazards are the lead dust produced during crushing and dry grinding operations, and lead fumes and lead oxide encountered in sintering, blast-furnace reduction and refining.

Lead sheet and pipe are used principally for the construction of equipment for storing and handling sulphuric acid. The use of lead for water and town gas pipes is limited nowadays. The hazards of working with lead increase with temperature. If lead is worked at temperatures below 500 °C, as in soldering, the risk of fume exposure is far less than in lead welding, where higher flame temperatures are used and the danger is higher. The spray coating of metals with molten lead is dangerous since it gives rise to dust and fumes at high temperatures (74).

The demolition of steel structures such as bridges and ships that have been painted with lead-based paints frequently gives rise to cases of lead poisoning. When metallic lead is heated to 550 °C, lead vapour will be evolved and will become oxidized. This is a condition that is liable to be present in metal refining, the melting of bronze and brass, the spraying of metallic lead, lead burning, chemical plant plumbing, ship breaking and the burning, cutting and welding of steel structures coated with paints containing lead tetroxide.

3.1.5.2 Absorption

The degree of absorption depends on the proportion of the dust accounted for by particles less than 5 microns in size and the exposed worker's respiratory minute volume. Since the most important route of lead absorption is by the lungs, the particle size of industrial lead dust is of considerable significance and this depends on the nature of the operation giving rise to the dust. Fine dust of respirable particle size is produced by processes such as the pulverizing and blending of lead colours, the abrasive working of lead-based fillers in automobile bodies and the dry rubbing-down of lead paint. The exhaust gases of gasoline engines yield lead chloride and lead bromide particles of 1 micron diameter. The larger particles, however, may be ingested and be absorbed via the stomach. A more informative picture of the hazard associated with a sample of lead dust might be given by including a size distribution as well as a total lead determination.

In the human body, inorganic lead is not metabolized but is directly absorbed, distributed and excreted. The rate at which lead is absorbed depends on its chemical and physical form and on the physiological characteristics of the exposed person such as nutritional status and age. Inhaled lead deposited in the lower respiratory tract is completely absorbed. The amount of lead absorbed from the gastrointestinal tract of adults is typically 10 to 15% of the ingested quantity; for pregnant women and children, the amount absorbed can increase to as much as 50%. The quantity absorbed increases significantly under fasting conditions and with iron or calcium deficiency (75).

Once in the blood, lead is distributed primarily among three compartments — blood, soft tissue (kidney, bone marrow, liver, and brain), and mineralizing tissue (bones and teeth). Mineralizing tissue contains about 95% of the total body burden of lead in adults.

The lead in mineralizing tissues accumulates in subcompartments that differ in the rate at which lead is resorbed. In bone, there is both a labile component, which readily exchanges lead with the blood, and an inert pool. The lead in the inert pool poses a special risk because it is a potential endogenous source of lead. When the body is under physiological stress such as pregnancy, lactation or chronic disease, this normally inert lead can be mobilized, increasing the lead level in blood. Because of these mobile lead stores, significant drops in a person's blood lead level can take several months or sometimes years, even after complete removal from the source of lead exposure.

Of the lead in the blood, 99% is associated with erythrocytes; the remaining 1% is in the plasma, where it is available for transport to the tissues. The blood lead not retained is either excreted by the kidneys or through biliary clearance into the gastrointestinal tract. In single-exposure studies with adults, lead has a half-life, in blood, of approximately 25 days; in soft tissue, about 40 days; and in the non-labile portion of bone, more than 25 years. Consequently, after a single exposure a person's blood lead level may begin to return to normal; the total body burden, however, may still be elevated (76).

For lead poisoning to develop, major acute exposures to lead need not occur. The body accumulates this metal over a lifetime and releases it slowly, so even small doses, over time, can cause lead poisoning. It is the total body burden of lead that is related to the risk of adverse effects.

3.1.5.3 Physiological effects

Whether lead enters the body through inhalation or ingestion, the biologic effects are the same; there is interference with normal cell function and with a number of physiological processes.

Neurological effects: The most sensitive target of lead poisoning is the nervous system. In children, neurological deficits have been documented at exposure levels once thought to cause no harmful effects. In addition to the lack of a precise threshold, childhood lead toxicity may have permanent effects. Some studies showed that damage to the central nervous system (CNS) that occurred as a result of lead exposure at age 2 resulted in continued deficits in neurological development, such as lower IQ scores and cognitive deficits, at age 5 (77).

Adults also experience CNS effects at relatively low blood lead levels, manifested by subtle behavioural changes, fatigue and impaired concentration. Peripheral nervous system damage, primarily motor, is seen mainly in adults. Lead neuropathy is believed to be a motor neuron, anterior horn cell disease with peripheral dying-back of the axons. Frank wrist drop occurs only as a late sign of lead intoxication.

Lead inhibits the body's ability to make hemoglobin by interfering with several enzymatic steps in the heme pathway. A decrease in the activity of ferrochelatase enzyme results in an increase of the substrate, erythrocyte protoporphyrin (EP), in the red blood cells. Recent data indicate that the EP level, which has been used to screen for lead toxicity in the past, is not sufficiently sensitive at lower levels of blood lead and is therefore not as useful a screening test for lead poisoning as previously thought.

Lead can induce two types of anemia. Acute high-level lead poisoning has been associated with hemolytic anemia. In chronic lead poisoning, lead induces anemia by both interfering with erythropoiesis and by diminishing red blood cell survival. It should be emphasized, however, that anemia is not an early manifestation of lead poisoning and is evident only when the blood lead level is significantly elevated for prolonged periods.

A strong inverse correlation exists between blood lead levels and levels of vitamin D. Because the vitamin D-endocrine system is responsible in large part for the maintenance of extra- and intra-cellular calcium homeostasis, it is likely that lead impairs cell growth and maturation and tooth and bone development.

A direct effect on the kidney of long-term lead exposure is nephropathy. There is also evidence of an association between lead exposure and hypertension, an effect that may be mediated through renal mechanisms. Maternal lead stores readily cross the placenta, placing the foetus at risk. Increasing evidence indicates that lead not only affects the viability of the foetus, but development as well. Developmental consequences of prenatal exposure to low levels of lead include reduced birth weight and premature birth. Lead is an animal teratogen; however, most studies in humans have failed to show a relationship between lead levels and congenital malformations.

Inorganic lead and inorganic lead compounds have been classified as Group 2B, possible human carcinogens, by the International Agency for Research on Cancer (IARC) (78).

3.1.5.4 Organic lead intoxication

The absorption of a sufficient quantity of tetraethyllead, whether briefly at a high rate or for prolonged periods at a lower rate, induces acute intoxication of the CNS. The milder manifestations are those of insomnia, lassitude and nervous excitation which reveal itself in lurid dreams and dream-like waking states of anxiety, in association with tremor, hyper-reflexia, spasmodic muscular contractions, bradycardia, vascular hypotension and hypothermia. The more severe responses include recurrent (sometimes nearly continuous) episodes of complete disorientation with hallucinations, facial contortions and intense general somatic muscular activity with resistance to physical restraint. Such episodes may be converted abruptly into maniacal or violent convulsive seizures which may terminate in coma and death.

3.1.5.5 Legislation

Clinical lead poisoning has historically been one of the most important occupational diseases, and it remains a major risk today. The considerable body of scientific knowledge concerning the toxic effects of lead has been enriched since the 1980s by significant new knowledge regarding the more subtle subclinical effects. Similarly, in a number of countries it was felt necessary to redraft or modernize work protective measures enacted over the last half-century and more (25).

Some regulation, such as the Occupational Safety and Health Administration (OSHA) lead standard, specifies the permissible exposure limit (PEL) of lead in the workplace, the frequency and extent of medical monitoring, and other responsibilities of the employer. As of this writing, if blood monitoring reveals a blood lead level greater than 40 µg/dL, the worker must be notified in writing and provided with medical examination. If a worker's blood lead level reaches 60 µg/dL (or averages 50 µg/dL or more), the employer is obligated to remove the employee from excessive exposure, with maintenance of seniority and pay, until the employee's blood lead level falls below 40 µg/dL (79).

3.1.6 Nickel

Since nickel, copper and iron occur as distinct minerals in the sulphide ores, mechanical methods of concentration, such as flotation and magnetic separation, are applied after the ore has been crushed and ground. The nickel concentrate is converted to nickel sulphide matte by roasting or sintering. The matte is refined by electrowinning or by the Mind process. In the Mind process, the matte is ground, calcined and treated with carbon monoxide at 50 °C to form gaseous nickel carbonyl ($Ni(CO)_4$), which is then decomposed at 200 to 250 °C to deposit pure nickel powder. Worldwide production of nickel is approximately 1.2 million ton/year (80).

More than 3,000 nickel alloys and compounds are commercially produced. Stainless steel and other Ni-Cr-Fe alloys are widely used for corrosion-resistant equipment, architectural applications and cooking utensils. Monel metal and other Ni-Cu alloys are used in coinage, food-processing machinery and dairy equipment. Ni-Al alloys are used for magnets and catalyst production. Ni-Cr alloys are used for heating elements, gas turbines and jet engines. Alloys of nickel with precious metals are used in jewellery. Nickel metal, its compounds and alloys have many other uses, including electroplating, magnetic tapes and computer components, arc-welding rods, surgical and dental prostheses, nickel-cadmium batteries, paint pigments (e.g., yellow nickel titanate), moulds for ceramic and glass containers, and catalysts for hydrogenation reactions, organic syntheses and the final methanation step of coal gasification. Occupational exposures to nickel also occur in recycling operations, since nickel-bearing materials, especially from the steel industry, are commonly melted, refined and used to prepare alloys similar in composition to those that entered the recycling process.

3.1.6.1 Toxicity

Human health hazards from occupational exposures to nickel compounds generally fall into three major categories: allergy, rhinitis, sinusitis and respiratory diseases, and cancers of the nasal cavities, lungs and other organs. The health effects of nickel were summarized in Table 2.

Nickel and nickel compounds are among the most common causes of allergic contact dermatitis. This problem is not limited to persons with occupational exposure to nickel compounds; dermal sensitization occurs in the general population from exposures to nickel-containing coins, jewellery, watch cases and clothing fasteners. In nickel-exposed persons, nickel dermatitis usually begins as a papular erythema of the hands. The skin gradually becomes eczematous, and, in the chronic stage, lichenification frequently develops. Nickel sensitization sometimes causes conjunctivitis, eosinophilic pneumonitis, and local or systemic reactions to nickel-containing implants (e.g., intraosseous pins, dental inlays, cardiac valve prostheses and pacemaker wires). Ingestion of nickel-contaminated tap water or nickel-rich foods can exacerbate hand eczema in nickel-sensitive persons.

Workers in nickel refineries and nickel electroplating shops, who are heavily exposed to inhalation of nickel dusts or aerosols of soluble nickel compounds, may develop chronic diseases of the upper respiratory tract. Chronic diseases of the lower respiratory tract including bronchitis, pulmonary fibrosis have also been reported, but such conditions are infrequent (81).

Epidemiological studies of nickel-refinery workers in Canada, Wales, Germany, Norway and Russia have documented increased mortality rates from cancers of the lung and nasal cavities. Certain groups of nickel-refinery workers have also been reported to have increased incidences of other malignant tumours, including carcinomas of the larynx, kidney, prostate or stomach, and sarcomas of soft tissues, but the statistical significance of these observations is questionable. The increased risks of cancers of the lungs and nasal cavities have occurred primarily among workers in refinery operations that entail high nickel exposures, including roasting, smelting and electrolysis. Although these cancer risks have generally been associated with exposures to insoluble nickel compounds, such as nickel subsulphide and nickel oxide, exposures to soluble nickel compounds have been implicated in electrolysis workers.

Epidemiological studies of cancer risks among workers in nickel-using industries have generally been negative, but recent evidence suggests slightly increased lung cancer risks among welders, grinders, electroplaters and battery makers. Such workers are often exposed to dusts and fumes that contain mixtures of carcinogenic metals (e.g., nickel and chromium, or nickel and cadmium). Based on an evaluation of epidemiological studies, the International Agency for Research on Cancer (IARC) concluded in 1990: "There is sufficient evidence in humans for the carcinogenicity of nickel sulphate and of the combinations of nickel sulphides and oxides encountered in the nickel refining industry. There is inadequate evidence in humans for the carcinogenicity of nickel and nickel alloys". Nickel compounds have been classified as carcinogenic to humans (Group 1), and metallic nickel as possibly carcinogenic to humans (Group 2B) (82).

3.1.6.2 Biological monitoring

Analyses of nickel concentrations in urine and serum samples may reflect the recent exposures of workers to metallic nickel and soluble nickel compounds, but these assays do not furnish reliable measures of the total body nickel burden. The uses and limitations of biological monitoring of nickel-exposed workers have been summarized. A technical report on analysis of nickel in body fluids was issued in 1994 by the Commission on Toxicology of the International Union of Pure and Applied Chemistry (IUPAC) (83). The National Maximum Workplace Concentration Committee (NMWCC) of the Netherlands proposed that urine nickel concentration 40 µg/g creatinine, or serum nickel concentration 5 µg/l (both measured in samples obtained at the end of a working week or a work shift) be considered warning limits for further investigation of workers exposed to nickel metal or soluble nickel compounds. If a biological monitoring programme is implemented, it should augment an environmental monitoring programme, so that biological data are not used as a surrogate for exposure estimates (83).

3.1.7 Mercury

3.1.7.1 Inorganic mercury

Mercury combines readily with sulphur and halogens at ordinary temperatures and forms amalgams with all metals except iron, nickel, cadmium, aluminium, cobalt and platinum. It reacts exothermically with alkaline metals, is attacked by nitric acid but not by hydrochloric acid and, when hot, will combine with sulphuric acid. Inorganic mercury is found in nature in the form of the sulphide (HgS) as cinnabar ore, which has an average mercury content of 0.1 to 4%. Mercury ore is extracted by underground mining, and mercury metal is separated from the ore by roasting in a rotary kiln or shaft furnace, or by reduction with iron or calcium oxide. The vapour is carried off in the combustion gases and is condensed in vertical tubes.

The most important uses of metallic mercury and its inorganic compounds have included the treatment of gold and silver ores; the manufacture of amalgams; the manufacture and repair of measurement or laboratory apparatus; the manufacture of incandescent electric bulbs, mercury vapour tubes, radio valves, x-ray tubes, switches, batteries, rectifiers, etc.; as a catalyst for the production of chlorine and alkali and the production of acetic acid and acetaldehyde from acetylene; chemical, physical and biological laboratory research; gold, silver, bronze and tin plating; tanning and currying; feltmaking; taxidermy; textile manufacture; photography and photogravure; mercury-based paints and pigments; and the manufacture of artificial silk (84). Some of these uses have been discontinued because of the toxic effects that the mercury exposure exerted upon workers.

3.1.7.2 Organic mercury compounds

Organic compounds of mercury may be considered as the organic compounds in which the mercury is chemically linked directly to a carbon atom. Carbon-mercury bonds have a wide range of stability; in general, the carbon-to-mercury bond in aliphatic compounds is more stable than that in aromatic compounds. It was estimated that more than 400 phenyl mercurials and at least that number of alkyl mercury compounds have been synthesized. The three most important groups in common usage are the alkyls, the aromatic hydrocarbons or aryls and the alkoxyalkyls. Examples of aryl mercury compounds are phenylmercuric acetate (PMA), nitrate, oleate, propionate and benzoate. Most available information is about PMA.

In medical practice, organic mercury compounds are used as antiseptics, germicides, diuretics and contraceptives. In the field of pesticides, they serve as algicides, fungicides, herbicides and slimacides, and as preservatives in paints, waxes and pastes; they are used for mildew suppression, in antifouling paints, in latex paints and in the fungus-proofing of fabrics, paper, cork, rubber and wood for use in humid climates. In the chemical industry, they act as catalysts in a number of reactions and the mercury alkyls are used as alkylating agents in organic syntheses.

3.1.7.3 Toxicity

Vapour inhalation is the main route for the entry of metallic mercury into the body. Around 80% of inhaled mercury vapour is absorbed in the lung (alveoli). Digestive absorption of metallic mercury is negligible (lower than 0.01% of the administered dose). The main routes of entry of inorganic mercury compounds (mercury salts) are the lungs (atomization of mercury salts) and the gastrointestinal tract. In the latter case, absorption is often the result of accidental or voluntary ingestion. It is estimated that 2 to 10% of ingested mercury salts are absorbed through the intestinal tract. The health effects of mercury were summarized in Table 2.

Skin absorption of metallic mercury and certain of its compounds is possible, although the rate of absorption is low. After entry into the body, metallic mercury continues to exist for a short time in metallic form, which explains its penetration of the blood-brain barrier. In blood and tissues metallic mercury is rapidly oxidized to Hg^{2+} mercury ion, which fixes to proteins. In the blood, inorganic mercury is also distributed between plasma and red blood cells. The kidney and brain are the sites of deposition following exposure to metallic mercury vapours, and the kidney following exposure to inorganic mercury salts (85).

3.1.7.4 Acute poisoning

The symptoms of acute poisoning include pulmonary irritation (chemical pneumonia), perhaps leading to acute pulmonary oedema. Renal involvement is also possible. Acute poisoning is more often the result of accidental or voluntary ingestion of a mercury salt. This leads to severe inflammation of the gastrointestinal tract followed rapidly by renal insufficiency due to necrosis of the proximal convoluted tubules.

3.1.7.5 Chronic exposure

Chronic mercury poisoning usually starts insidiously, which makes the early detection of incipient poisoning difficult. The main target organ is the nervous system. Initially, suitable tests can be used to detect psychomotor and neuro-muscular changes and slight tremor. Slight renal involvement (proteinuria, albuminuria, enzymuria) may be detectable earlier than neurological involvement. If excessive exposure is not corrected, neurological and other manifestations (e.g., tremor, sweating, dermatography) become more pronounced, associated with changes in behaviour and personality disorders and, perhaps, digestive disorders (stomatitis, diarrhoea) and a deterioration in general status (anorexia, weight loss). Once this stage has been reached, termination of exposure may not lead to total recovery.

In chronic mercury poisoning, digestive and nervous symptoms predominate and, although the former are of earlier onset, the latter are more obvious; other significant but less intense symptoms may be present. The duration of the period of mercury absorption preceding the appearance of clinical symptoms depends on the level of absorption and individual factors. The main early signs include slight digestive disorders, in particular, loss of appetite; intermittent tremor, sometimes in specific muscle groups; and neurotic disorders varying in intensity. The course of intoxication may vary considerably from case to case. If exposure is terminated immediately upon the appearance of the first symptoms, full recovery usually occurs; however, if exposure is not terminated and the intoxication becomes firmly established, no more than an alleviation of symptoms can be expected in the majority of cases.

There have been studies over the years on the relationships between renal function and urinary mercury levels. The effects of low-level exposures are still not well documented or understood. At higher levels (above 50 µg/g (micrograms per gram) abnormal renal function (as evidenced by N-acetyl-B-D-glucosaminidase (NAG), which is a sensitive indicator of damage to the kidneys) have been observed. The NAG levels were correlated with both the urinary mercury levels and the results of neurological and behavioural testing (86).

Chronic poisoning is accompanied by mild anemia sometimes preceded by polycythaemia resulting from bone marrow irritation. Lymphocytosis and eosinophilia have also been observed.

Absorption of phenylmercuric acetate (PMA) may occur through inhalation of aerosols containing PMA, through skin absorption or by ingestion. The solubility of the mercurial and the particle size of the aerosols are determining factors for the extent of absorption. PMA is more efficiently absorbed by ingestion than are inorganic mercuric salts. Phenylmercury is transported mainly in blood and distributed in the blood cells (90%), accumulates in the liver and is there decomposed into inorganic mercury. Some phenylmercury is excreted in the bile. The main portion absorbed in the body is distributed in the tissues as inorganic mercury and accumulated in the kidney. On chronic exposure, mercury distribution and excretion follow the pattern seen on exposure to inorganic mercury.

Occupational exposure to phenylmercury compounds occurs in the manufacture and handling of products treated with fungicides containing phenylmercury compounds. Acute inhalation of large amounts may cause lung damage. Exposure of the skin to a concentrated solution of phenylmercury compounds may cause chemical burns with blistering. Ingestion of large amounts of phenylmercury may cause renal and liver damage. Chronic poisoning gives rise to renal damage due to accumulation of inorganic mercury in the renal tubules.

Available clinical data do not permit extensive conclusions about dose-response relationships. They suggest, however, that phenylmercury compounds are less toxic than inorganic mercury compounds or long-term exposure. There is some evidence of mild adverse effects on the blood.

Alkyl mercury compounds. From a practical point of view, the short-chained alkyl mercury compounds, like methylmercury and ethylmercury, are the most important, although some exotic mercury compounds, generally used in laboratory research, have led to spectacular rapid deaths from acute poisoning. These compounds have been extensively used in seed treatment where they have been responsible for a number of fatalities. Methylmercuric chloride forms white crystals with a characteristic odour, while ethylmercury chloride; (chloroethylmercury) forms white flakes. Volatile methylmercury compounds, like methylmercury chloride, are absorbed to about 80% upon inhalation of vapour. More than 95% of short-chained alkyl mercury compounds are absorbed by ingestion, although the absorption of methylmercury compounds by the skin can be efficient, depending on their solubility and concentration and the condition of the skin (87).

Methylmercury is transported in the red blood cells (95%), and a small fraction is bound to plasma proteins. The distribution to the different tissues of the body is rather slow and it takes about four days before equilibrium is obtained. Methylmercury is concentrated in the central nervous system and especially in grey matter. About 10% of the body burden of mercury is found in the brain. The highest concentration is found in the occipital cortex and the cerebellum. In pregnant women methylmercury is transferred in the placenta to the foetus and especially accumulated in the fetal brain.

3.1.7.6 Toxicity of organic mercury

Poisoning by alkyl mercury may occur on inhalation of vapour and dust containing alkyl mercury and in the manufacture of the mercurial or in handling the final material. Skin contact with concentrated solutions results in chemical burns and blistering. In small agricultural operations there is a risk of exchange between treated seed and products intended for food, followed by involuntary intake of large amounts of alkyl mercury. On acute exposure the signs and symptoms of poisoning have an insidious onset and appear with a latency period which may vary from one to several weeks.

On chronic exposure the onset is more insidious, but the symptoms and signs are essentially the same, due to the accumulation of mercury in the central nervous system, causing neuron damage in the sensory cortex, such as visual cortex, auditory cortex and the pre- and post-central areas. The signs are characterized by sensory disturbances with paresthaesia in the distal extremities, in the tongue and around the lips. With more severe intoxications ataxia, concentric constrictions of the visual fields, impairment of hearing and extrapyramidal symptoms may appear. In severe cases chronic seizures occur.

The period in life most sensitive to methylmercury poisoning is the time in utero; the foetus seems to be between 2 and 5 times more sensitive than the adult. Exposure in utero results in cerebral palsy, partly due to inhibition of the migration of neurons from central parts to the peripheral cortical areas. In less severe cases retardation in the psychomotor development has been observed.

The most common alkoxyalkyl compounds used are methoxyethyl mercury salts (e.g., methoxyethylmercury acetate), which have replaced the short-chain alkyl compounds in seed treatment in many industrial countries, in which the alkyl compounds have been banned due to their hazardousness. The available information is very limited. Alkoxyalkyl compounds are absorbed by inhalation and by ingestion more efficiently than inorganic mercury salts (88). The distribution and excretion patterns of absorbed mercury follow those of inorganic mercury salts. Excretion occurs through the intestinal tract and the kidney. To what extent unchanged alkoxyalkyl mercury is excreted in humans is unknown. Exposure to alkoxyalkyl compounds can occur in the manufacture of the compound and in handling the final product(s) treated with the mercurial. Methoxyethyl mercury acetate is a vesicant when applied in concentrated solutions to the skin. Inhalation of methoxyethyl mercury salt dust may cause lung damage, and chronic poisoning due to long-term exposure may give rise to renal damage.

Most exposure to organic mercury compounds involves mixed exposure to mercury vapour and the organic compound, as the organic mercury compounds decompose and release mercury vapour. All technical measures pertaining to exposure to mercury vapour should be applied for exposure to organic mercury compounds. Thus, contamination of clothes and/or parts of the body should be avoided, as it may be a dangerous source of mercury vapour close to the breathing zone. Special protective work clothes should be used and changed after the workshift. Spray painting with paint containing mercurials requires respiratory protective equipment and adequate ventilation. The short-chained alkyl mercury compounds should be eliminated and replaced whenever possible. If handling cannot be avoided, an enclosed system should be used, combined with adequate ventilation, to limit exposure to a minimum.

Great care must be exercised in preventing the contamination of water sources with mercury effluent since the mercury can be incorporated into the food chain, leading to disasters such as that which occurred in Minamata, Japan.

3.1.8 Metal carbonyls

Metal carbonyls have the general formula $Me_x(CO)_y$, and are formed by combination of the metal (Me) with carbon monoxide (CO). Physical properties of some metal carbonyls are listed in Table 3 (89). Most are solids at ordinary temperatures, but nickel carbonyl, iron pentacarbonyl and ruthenium pentacarbonyl are liquids, and cobalt hydrocarbonyl is a gas. Since iron pentacarbonyl and cobalt hydrocarbonyl also have high vapour pressures and potential for inadvertant formation, they warrant serious consideration as possible occupational toxicants. Most metal carbonyls react vigorously with oxygen and oxidizing substances, and some ignite spontaneously. Upon exposure to air and light, nickel carbonyl decomposes to carbon monoxide and particulate nickel metal, cobalt hydrocarbonyl decomposes to cobalt octacarbonyl and hydrogen, and iron pentacarbonyl decomposes to iron nonacarbonyl and carbon monoxide (90).

Metal carbonyls are used in isolating certain metals (e.g., nickel) from complex ores, for producing carbon steel, and for metallizing by vapour deposition. They are also used as catalysts in organic reactions (e.g., cobalt hydrocarbonyl or nickel carbonyl in olefin oxidation; cobalt octacarbonyl for the synthesis of aldehydes; nickel carbonyl for the synthesis of acrylic esters). Iron pentacarbonyl is used as a catalyst for various organic reactions, and is decomposed to make finely powdered, ultra pure iron (so-called carbonyl iron), which is used in the computer and electronics industries. Methycyclopentadienyl manganese tricarbonyl (MMT) $(CH_3C_5H_4Mn(CO)_3)$ is an antiknock additive to gasoline.

Metal carbonyl	Mol. Wt.	Sp. Gr. (20°C)	M.P. (°C)	B.P. (°C)	V.P. (25°C) mm Hg
$Ni(CO)_4$	170.75	1.31	–19	43	390
$CoH(CO)_4$	171.99	–	–26	–	high
$Co_2(CO)_8$	341.95	1.87	51	52*	1.5
$Co_4(CO)_{12}$	571.86	–	60*	–	very low
$Cr(CO)_6$	220.06	1.77	110*	151	0.4
$Fe_2(CO)_9$	363.79	2.08	80*	–	–
$Fe(CO)_5$	195.90	1.46	–25	103	30.5
$Fe(CO)_4$	167.89	2.00	approx. 140*	–	–
$Mo(CO)_6$	264.00	1.96	150*	156	0.2
$Ru(CO)_5$	241.12	–	–22	–	–
$W(CO)_6$	351.91	2.65	approx. 150*	175	0.1

*Decomposition starts at temperature shown.

Table 3. Physical properties of some metal carbonyls.

3.1.8.1 Toxicity

The toxicity of a given metal carbonyl depends on the toxicity of carbon monoxide and of the metal from which it is derived, as well as the volatility and instability of the carbonyl itself. The principal route of exposure is inhalation, but skin absorption can occur with the liquid carbonyls. The relative acute toxicity (LD_{50} for the rat) of nickel carbonyl, cobalt hydrocarbonyl and iron pentacarbonyl may be expressed by the ratio 1:0.52:0.33. Inhalation exposures of experimental animals to these substances induce acute interstitial pneumonitis, with pulmonary oedema and capillary damage, as well as injury to the brain, liver and kidneys (91).

Iron pentacarbonyl can be formed inadvertently when carbon monoxide, or a gas mixture containing carbon monoxide, is stored under pressure in steel cylinders or fed through steel pipes, when illuminating gas is produced by petroleum reforming, or when gas welding is carried out. Presence of carbon monoxide in emission discharges from blast furnaces, electric arc furnaces and cupola furnaces during steel-making can also lead to the formation of iron pentacarbonyl.

3.1.8.2 Nickel carbonyl

Nickel carbonyl ($Ni(CO)_4$) is mainly used as an intermediate in the Mind process for nickel refining, but it is also used for vapour-plating in the metallurgical and electronics industries and as a catalyst for synthesis of acrylic monomers in the plastics industry. Inadvertent formation of nickel carbonyl can occur in industrial processes that use nickel catalysts, such as coal gasification, petroleum refining and hydrogenation reactions, or during incineration of nickel-coated papers that are used for pressure-sensitive business forms (92).

3.1.8.3 Toxicity

Acute, accidental exposure of workers to inhalation of nickel carbonyl usually produces mild, non-specific, immediate symptoms, including nausea, vertigo, headache, dyspnoea and chest pain. These initial symptoms usually disappear within a few hours. After 12 to 36 hours, and occasionally as long as 5 days after exposure, severe pulmonary symptoms develop, with cough, dyspnoea, tachycardia, cyanosis, profound weakness and often gastrointestinal symptoms. Human fatalities have occurred 4 to 13 days after exposure to nickel carbonyl; deaths have resulted from diffuse interstitial pneumonitis, cerebral hemorrhage or cerebral oedema. In addition to pathologic lesions in the lungs and brain, lesions have been found in liver, kidneys, adrenals and spleen. In patients who survive acute nickel carbonyl poisoning, pulmonary insufficiency often causes protracted convalescence. Nickel carbonyl is carcinogenic and teratogenic in rats; the European Union has classified nickel carbonyl as an animal teratogen. Processes that use nickel carbonyl constitute disaster hazards, since fire and explosion can occur when nickel carbonyl is exposed to air, heat, flames or oxidizers. Decomposition of nickel carbonyl is attended by additional toxic hazards from inhalation of its decomposition products, carbon monoxide and finely particulate nickel metal.

Chronic exposure of workers to inhalation of low atmospheric concentrations of nickel carbonyl (0.007 to 0.52 mg/m^3) can cause neurological symptoms such as insomnia, headache, dizziness, memory loss, and other manifestations including chest tightness, excessive sweating, alopecia. Electroencephalographic abnormalities and elevated serum monoamine oxidase activity have been observed in workers with chronic exposures to nickel carbonyl. A synergistic effect of cigarette smoking and nickel carbonyl exposure on the frequency of sister-chromatid exchanges was noted in a cytogenetic evaluation of workers with chronic exposure to nickel carbonyl.

Because of its flammability and tendency to explode, nickel carbonyl should be stored in tightly closed containers in a cool, well-ventilated area, away from heat and oxidizers such as nitric acid and chlorine. Flames and sources of ignition should be prohibited wherever nickel carbonyl is handled, used or stored. Nickel carbonyl should be transported in steel cylinders. Foam, dry chemical, or CO_2 fire extinguishers should be used to extinguish burning nickel carbonyl, rather than a stream of water, which might scatter and spread the fire.

Exposures are classified as "mild" if the initial 8-h specimen of urine has a nickel concentration less than 100 µg/l, "moderate" if the nickel concentration is 100 to 500 µg/l, and "severe" if the nickel concentration exceeds 500 µg/l (93). Sodium diethyldithiocarbamate is the drug of choice for chelation therapy of acute nickel carbonyl poisoning. Ancillary therapeutic measures include bed rest, oxygen therapy, corticosteroids and prophylactic antibiotics. Carbon monoxide poisoning may occur simultaneously and requires treatment.

4. Comparison of metal concentrations in air samples

In literature, the most published articles on metal concentrations in air phases is related with lead, nickel, cadmium and arsenic levels in airborne aerosol samples. Liang et al. (1990) examined six metal levels in air taken from laboratory and clean room (32) Gucer et al.

(1992) determined six metal concentrations in ambient air taken from Malatya city depending on month of year and distance from center of city (28). Jaradat and momani (1999) analyzed soil, plant and air samples taken from both sides of the major highway connecting for Cu, Pb, Cd and Zn (31). Fernandez et al. (2000) determined five metal concentrations in four fractionation of particulate matters in air (94). Fuchtjohann et al. (2000) used GFAAS and ICP-MS to determine soluble ($NiCl_2$ and partly $NiCO_3$) and insoluble (NiO) nickel compounds in ambient air dusts (airborne particulate matters) taken from locations close to two metallurgical plants (95). They found maximum concentrations in total nickel at both sampling sites as 40 ng/m^3 and 160 ng/m^3 whereas the mean values were 9 ng/m^3 and 28 ng/m^3 (95). Bolt et al. (2000) determined Ni levels in airborne dusts collected from a metal factory processing nickel and nickel alloys (96). They found Ni concentrations in range of 10.000-4.920.000 ng/m^3. Hadad et al. (2003) found Pb, Cr and Fe concentrations in Tehran air samples as 1.040.000, 56.000 and 2.720.000 ng/m^3, respectively (30). Wada et al. (2001) reported mean Pb, Ni and Mn concentrations in airborne samples taken from Nagasaki as 8, 4 and 12 ng/m^3 (37). Bhat and Pillai (1997) determined mean 0.42 ng Be /m3 in air sample taken from the vicinity of Be metal plant at New Bombay (97). Gurjar and Mohan (2003) reported Cd, Cr and Ni concentrations in atmospheric environment at fourten cities of India in ranges of 1-21, 16-207 and 23-257 ng/m^3, respectively (34). Pekney and Davidson (2005) determined 28 metal concentrations in ambient particulate matter by ICP-MS (33). Vijayanand et al. (2008) assessed seven trace metals in the ambient air of Coimbatore city in India (29). They found Ni, Pb and Cr concentrations in range of BDL-310, 210-620 and 5-880 ng/m^3, respectively (29). Limbeck et al. (2009) found Ni, Pb and Cr concentrations in ranges of 6-10, 9-11 and 4-6 ng/m^3, respectively (27). Canepari et al. (2009) compared XRF, and ICP-OES results for multielement concentrations in ambient air suspended particulate matter (98). Chen and Lippmann (2009) reviewed effect of metal concentrations in ambient air particulate matter on human health (35). They reported that Ni, Pb and Cr levels in PM2.5 for 13 city in USA were in ranges of 1-2, 2-14 and 1-3 ng/m^3, respectively. Morishita et al. (2006) attempted to determine source of pollution by using concentrations of trace metals including Pb, Zn, Cd and Fe in PM2.5 (99). Odabasi et al. (2002) determined 11 trace metals in ambient air samples taken from Izmir city center and compared their results with other values reported in literature (36). Newhook et al. (2003) reviewed trace metal concentrations in ambient air (PM10) samples near copper smelter and refineries, and zinc plants in Canada (40). Mohanraj et al. (2004) determined six trace metals in airborne samples from India (38). They found 2.147 ng Pb /m3 in particulate matter taken from industrial location. Pierre et al. (2002) attempted to determine relationship between blood lead concentration of workers employed in crystal industry and ambient air lead (100). Vanhoof et al. (2003) determined Pb, Cu and Zn concentrations in ambient air from nonferrous metal industry (39). Fang et al. reviewed 7 metal concentrations in particulate matter (PM2.5 and PM10) taken fro Asia countries between 2000-2004 (101). Krzemińska-Flowers et al. (2006) determined 15 trace metals in urban air particulate matter by ICP-MS (102). They compared the results taking into consideration summer and winter season. They found that winter-Ag, As and Hg concentrations in both PM3 and PM10 taken from all three locations having different pollution were higher than in summer season whereas other trace metals changed depending on season and locations. These results reveal the difficulties in comparison of trace metal concentrations in air samples due to many factors affected the values. The detailed concentrations and information about those studies were given in Table 4.

Location	Character	Size	Cd	Pb	Ni	Cu	As	Fe	Cr	Ref.
Boston-USA	PM2.5=16.5 µg/m³	PM2.5		240	9	11		62		7
St. Louis-USA	PM2.5=19.2 µg/m³			213	2	30		144		
Knoxville-USA	PM2.5=21.1 µg/m³			109	1	13		117		
Madison-USA	PM2.5=11.3 µg/m³			33	0.5	6		44		
Steubenville-USA	PM2.5=30.5 µg/m³			185	4	12		542		
Topeka-USA	PM2.5=12.2 µg/m³			72	0.6	7		72		
Schafberg	Out of City-Viena	PM10	0.4	8.9	7.0	7.2	0.7	250	3.8	27
Kendlerstrabe	City Suburb-Moderate Traffic		0.5	11	5.7	21	0.9	780	5.5	
Rinnböckstrabe	Iner-City-Density Traffic		0.5	11	9.9	20	1.2	740	5.0	
Malatya-Turkey	City Center		ND	3630-27270		1080-1760		4280-6030		28
Coimbatore-India	Residential and Traffic Area	SPM	BDL	340	120	660		2850	BDL	29
Coimbatore-India	Residential and Traffic Area		BDL	560	150	610		3150	470	
Coimbatore-India	Industrial Area		BDL	280	90	880		1850	630	
Coimbatore-India	Industrial Area		BDL	320	230	510		3650	170	
Coimbatore-India	Industrial Area		BDL	230	82	310		3300	430	
Coimbatore-India	Industrial and Traffic Area		BDL	520	BDL	290		2200	40	
Coimbatore-India	Industrial and Traffic Area		BDL	430	130	460		2950	390	
Coimbatore-India	Residential and Traffic Area		BDL	170	120	730		2200	280	
Coimbatore-India	Industrial and Traffic Area		BDL	320	100	690		4100	460	
Coimbatore-India	Industrial and Traffic Area		BDL	430	120	530		6000	510	
Tehran	Traffic-Industrial Summer	SPM		1.040.000				2720.000	56	30

Location	Character	Size	Cd	Pb	Ni	Cu	As	Fe	Cr	Ref.
	Traffic-Industrial Winter			1.000.000				1740.000	40	
Shiraz	Traffic-Industrial Summer			410				3252.000	15	
	Traffic-Industrial Winter			669				1891.000	15	
Amman-Jordan	Highweigh-Roadside			260-1370		260-600				31
Connecticut-laboratory	Nov. 1988				1.25	1.89		6.65		32
	June 1989				1.24	1.43		16.9		
Connecticut-Clean room	Nov 1988				0.19	0.31		1.54		
	June 1989				0.18	0.41		4.13		
Pitsburgh	Ambient particulate matter	PM2.5		20			20	200		33
Andhra Pradesh – India	Ambient air		14			68			32	34
Bihar			11			208			133	
Chandigarh (UT) - India			5			257			207	
Gujarat-India			16			45			51	
Haryana-India			8			134			90	
Himachal Pradesh-India			8			66			64	
Karnataka-India			1			43			17	
Kerala-India			6			23			18	
Orissa-India			8			78			144	
Punjab-India			4			107			82	
Rajasthan-India			20			63			16	
Tamil Nadu-India			6			28			17	
Uttar Pradesh-India			21			222			71	
Burlington-USA	Air particulate matter	PM2.5	-	-	2	2	-	-	2	35
Philadelphia-USA			-	5	6	4	-	-	2	
Atlanta-USA			-	3	-	2	1	1	-	
Detroit-USA			-	6	2	6	2	-	2	
Chicago-USA			-	6	1	4	1	1	1	
ST.Louis-USA			-	14	2	14	2	1	2	
Houston-USA			-	2	2	3	1	1	1	
Minneapolis-USA			-	5	2	3	2	-	2	
Boulder-USA			-	5	1	4	-	-	2	
Phoenix-USA			-	-	3	6	2	3	2	
Seattle-USA			-	4	2	3	1	1	2	
Sacramento-USA			-	-	10	6	2	-	2	

Location	Character	Size	Cd	Pb	Ni	Cu	As	Fe	Cr	Ref.
Riverside-Rubidojx-USA			-	6	2	6	2	-	3	
Izmir, Turkey	City suburb	TSP	8	111	39	154			11	36
Nagasaki City	High Traffic			7.95	3.78					37
India, (6 sampling station)	Urban residential, industrial, highway,		2.8	143.5	31.37	388.6			14.2	38
Location 1	Southern side the road	Rural		29		20				39
Location 2	Eastern side the road	PM10		17		12				
Location 3	Northern side the road	TSP		26		17				
Canada	Copper smelters	PM	1	197			28			40
Canada	Noranda-horne	PM	2	18			33-255			
Canada	Copper refineries	PM	0	34			8			
Canada	Zinc plants	PM	19							
Japon, Sapporo	Urban City	TSP		43.9	3.81	20.9		625	2.61	41
Japon, Tokyo				125	5.63	30.2		677	6.09	
China, Hong Kong	Airborne	TSP		79	-	88		1421	-	42
	Traffic	PM10		98740	8620	35380		860	6850	
		PM2.5		76860	5340	17320		250	2430	
China, Hong Kong	Industry	PM10		100520	9580	63530		790	5750	43
		PM2.5		91620	6000	36780		480	4510	
	Urban	PM10		62750	8270	15330		620	4970	
		PM2.5		60130	6330	9710		190	4190	
China, Shanghai	University	PM2.5		270	-	-		820	-	44
	Urban	PM2.5		280	-	-		900	-	
Vietnam, Ho Chi Minh	Urban	PM2		73	-	3		1222	-	45
		PM2-10		79	-	2		261	-	
		TSP		146	-	-		2904	-	
India,Sakinaka, Mumbai	Traffic Junction	SPM		1060	-	370		165500	-	46
India, Gandhinagar		SPM		820	-	1550		265500	-	
Indonesia, Bukit Tinggi	Tropical Jungle	PM2.5		1.22	-	<0.14		2.6	-	47
		PM2.5-10		<0.3	-	<0.16		14.8	-	
Indonesia, Pontianak	Rural	PM2.5		8.7	-	0.22		4.7	-	
		PM2.5-10		4.2	-	0.56		53	-	
		PM2.5		26	11	-		581	-	
Indonesia	Rural	PM2.5-10		3	2	-		1479	-	48
		TSP		39	18	-		2700	-	
		<2.5		88	54	32		994	360	

Location	Character	Size	Cd	Pb	Ni	Cu	As	Fe	Cr	Ref.
Indonesia	Temple	PM2.5-10		120	73	14		568	147	49
Taiwan	Traffic Junction (daytime)	TSP		180	-	240		1710	-	50
	Traffic Junction (nighttime)	TSP		180	-	230		1660	-	
	Inland Urban	PM10		150	30	-		1730	-	
Taiwan, Kaoshiung	Inland Industrial	PM10		80	90	-		2090	-	51
	Coastal Industrial	PM10		190	40	-		2140	-	
	Coastal Urban	PM10		340	30	-		1740	-	
Korea, Taejon	Industrial	TSP		269	33.6	54.9		1839	31.8	52
		PM10		195	42.6	32.4		1577	39.3	
Korea, Seoul	Urban	PM2.5		96.4	19.6	27.8		743	13.7	53
		PM10		124	47.8	50.1		2321	18.8	
Annual means-Sites in Europe	Rural	TSP	0.03-0.7		0.1-3.5		0.19-4.2			54
	Urban		0.11-1.2		1.6-13		0.8-3.1			
	Traffic		0.21-2.4		2.4-21		0.05-4.1			
	Industrial		0.2-23.7		2.2-102		1.2-97			

PM: particulate matter; TSP: Total Suspended Particles; SPM: Suspected Particule Matter
ND: Not detection; BDL: Below detection limit

Table 4. Reported metal concentrations and information related with air particules in literature. The metal concenttrations are ng/m³.

5. Conclusion

Exposure to chemicals is a serious public health problems that affect wildlife, soils, water, and air and can have very harmful human health effects. Exposures to chemicals including metals must be identified promptly, and individuals exposed to them must be evaluated and managed without delay. Major sources of metal emissions into environment can be able to change due to the updating in industrial activities and legislations. Before forbidding tetraethyl Pb in gasoline, automobile emissions were the primary source of Pb emissions while, nowadays, piston engine aircraft and industrial sources seem the two largest sources, depending on developing of countries (25).

As it is seen from Table 4, the reported toxic metal concentrations in air phases are significantly difference even if taking consideration their character such as rural, urban, traffic and industry. It was found that metal concentrations (ng/m³) were in the ranges of 0.03-21 for Cd, 0.18-1.040.000 for Pb, 0.1-9.580 for Ni, 0.14-63.530 for Cu, 0.7-255 for As, 1-3.252.000 for Fe and 1-6.850 for Cr. Particularly, the observed metal concentrations in air of industry area in China are extremely higher. As a result, there are a greet need for determinations of metal concentrations in air phases, to obtain reliable and considerable values.

6. References

[1] Goyer R., 2004, Issue Paper On The Human Health Effects Of Metals, U.S. Environmental Protection Agency, Lexington, MA, USA.

[2] Jarup L., Akesson A., 2009, Current status of cadmium as an environmental health problem, Toxicology and Applied Pharmacology, 238, 201–208

[3] Mamtani R., Stren P., Dawood I., Cheema S., 2011, Review Article-Metals and Disease: A Global Primary Health Care Perspective, Journal of Toxicology, 1-11 doi:10.1155/2011/319136

[4] Yaman M., 2006, Comprehensive comparison of trace metal concentrations in cancerous and non-cancerous human tissues, Current. Medicinal Chemistry, 13, 2513-2525.

[5] Turco RP., 1997, Effect of exposure to pollution, Earth under Siege; from air pollution to global change, Oxford University Press, Oxford, 191-192.

[6] Pope CA., Dockery DW., Schwartz J., 1995, Review Of Epidemiological Evidence Of Health-Effects Of Particulate Air-Pollution, Inhalation Toxicology, 7(1), 1-18.

[7] Laden F., Neas LM., Dockery DW., Schwartz J., 2000, Association of Fine Particulate Matter from Different Sources with Daily Mortality in Six U.S. Cities, Environmental Health Perspectives,108(10), 941-947.

[8] Kennedy IM., 2007, The health effects of combustion-generated aerosols, Proceedings of the Combustion Institute 31, 2757–2770

[9] Rovira J., Mari M., Nadal M., Schuhmacher M., Domingo JL., 2011, Levels of metals and PCDD/Fs in the vicinity of a cement plant: Assessment of human health risks, Journal of Environmental Science and Health, Part A, 46, 1075–1084.

[10] Mulgrew A. and Williams P., 2000, Biomonitoring of air quality using plants Air Hygiene Report No 10, Berlin, Germany, WHO CC, 165 pp.

[11] Chillrud S. N., David E ., Ross J. M., Sax S. N., Pederson D., Spengler J . D and Kin ney P. L., 2004, Elevated Airborne Exposures of Teenagers to Manganese, Chromium, and Iron from Steel Dust and New York City's Subway System, Environ. Sci. Technol., 38, 732-737.

[12] Wolterbeek B., 2002, Biomonitoring of trace element air pollution: principles, possibilities and perspectives, Environmental Pollution, 120, 11–21.

[13] Yaman, M., 2000, Nickel Speciation in Soil and the Relationship with Its Concentration in Fruits. Bulletin of Environmental Contamination and Toxicology, 65, 545-552.

[14] Yaman, M., 1997, Determination of Manganese in Vegetables By Atomic Absorption Spectrometry with Enrichment using Activated Carbon. Chemia Analityczna (Warsaw), 42(1), 79-86.

[15] Kaya G. and Yaman M., 2008, Online preconcentration for the determination of lead, cadmium and copper by slotted tube atom trap (STAT)-flame atomic absorption spectrometry, Talanta, 75, 1127-1133.

[16] Kaya G. and Yaman M., 2008, Trace metal concentrations in cupressecea leaves as biomonitors of environmental pollution, Trace Element and Electrolytes, 25(3), 156-164.

[17] Karaaslan N. M. and Yaman M., 2011, Use of STAT-Flame AAS in the Determination of Lead and Cadmium Pollution in Urban Center, Atom. Spectrosc., 32(4), 152-159.

[18] Kaya G., Ozcan C. and Yaman M., 2010, Flame Atomic Absorption Spectrometric Determination of Pb, Cd, and Cu in Pinus nigra L. and Eriobotrya japonica Leaves Used as Biomonitors in Environmental Pollution, Bull Environ. Contam. Toxicol., 84(2), 191-196.

[19] The Convention on Long-range Transboundary Air Pollution (CLRTAP), Geneva on 13 November 1979.

[20] United Nations Economic Commission for Europe (UN/ECE): 2000-01-18 from the EMEP program (Co-operative Programme for emissions of air pollutants through legally binding protocols).

[21] EIS, 2000. EU/UN-ECE: Commission proposes ratification of protocol on heavy metals. Europe Environment. 566, 9-10. Europe Information Service.

[22] NEPC, 1998. National Environment Protection Measure for Ambient Air Quality, National Environmental Protection Council, Commonwealth Government of Australia, 26 June.

[23] EA, 2000. State of Knowledge Report: Air Toxics and Indoor Air Quality in Australia (Draft), Environment Australia, Commonwealth of Australia, Canberra.

[24] US EPA, 1999, National Air Toxics Program: The Integrated Urban Strategy, United States Environmental Protection Agency's Federal Register Vol. 64, No. 137,1999'), available at: http://www.epa.gov/ttn/uatw/urban/urbanpg.html_.

[25] Cho SH., Richmond-Bryant J., Thornburg J., Portzer J., Vanderpool R., Cavender K., Rice J., 2011, A literature review of concentrations and size distributions of ambient airborne Pb-containing particulate matter, Atmospheric Environment, 45, 5005-5015.

[26] Cornellis R., 2006, Elemental speciation in human health risk assessment / WHO-Environmental Health Criteria 234. Toronto, Canada.

[27] Limbeck A., Handler M., Puls C., Zbiral J., Bauer H. and Puxbaum H., Handler M., 2009, Impact of mineral components and selected trace metals on ambient PM10 concentrations, Atmospheric Environment, 43 (3), 530-538.

[28] Gucer S., Demir M. Karagozler A.E. and Karakaplan M., 1992, Atmospheric distribution of some trace metals in Malatya, NATO ASI series, Vol: G 31, Industrial air pollution assessment and control, Springer-Verlag, Berlin.

[29] Vijayanand C., Rajaguru R., Kalaiselvi K., Selvam Panneer K. and Palanivel M., 2008, Assessment of heavy metal contents in the ambient air of the Coimbatore city, Tamilnadu, India, Journal of Hazardous Materials, 160 (2-3), 548-553.

[30] Hadad K., Mehdizadeh S. and Sohrabpour M., 2003, Impact of different pollutant sources on Shiraz air pollution using SPM elemental analysis, Environment International, 29 (1), 39-43.

[31] Jaradat Q. M. and Momani K. A., 1999, Contamination of roadside soil, plants, and air with heavy metals in Jordan, a comparative study, Turk J. Chem., 23, 209-220.

[32] Liang Z., Wei G-T, Irwin R. L., Walton A. P. and Michel R. G., 1990, Determination of subnanogram per cubic meter concentrations of metals in the air of a trace metal clean room by impaction graphite furnace atomic absorption and laser excited atomic fluorescence spectrometry, Anal. Chem., 62, 1452-1457.

[33] Pekney N. J. and Davidson C. I. 2005, Determination of trace elements in ambient aerosol samples, Analytica Chimica Acta, 540, 269-277.

[34] Gurjar B. R. and Mohan M., 2003, Potential Health Risks Due To Toxic Contamination In The Ambient Environment of Certain Indian States, Environmental Monitoring and Assessment, 82, 203-223.

[35] Chen L. C. and Lippmann M., 2009, Effects of Metals within Ambient Air Particulate Matter (PM) on Human Health, Inhalation Toxicology, 21, 1-31.

[36] Odabasi M., Muezzinoglu A. and Bozlaker A., 2002, Ambient Concentrations and Dry Deposition Fluxes of Trace Elements In Izmir, Turkey, Atmospheric Environment 36, 5841-5851.

[37] Wada M., Kido H., Kishikawa N., Tou T., Tanaka M., Tsubokura J., Shironita M., Matsui M., Kuroda N. and Nakashima K., 2001, Assesment of Air Pollution in Nagasaki City: Determination of Polycyclic Aromatic Hydrocarbons and Their Nitrated Derivatives, and Some Metals, Environmental Pollution, 115, 139-147.

[38] Mohanraj R., Azeez P. A. and Priscilla T., 2004, Heavy Metals in Airborne Particulate Matter of Urban Coimbatore, Arch. Environ. Contam. Toxicol. 47, 162–167.

[39] Vanhoof C., Chen H., Berghmans P., Corthouts V., De Brucker N. and Tirez K., 2003, A risk assessment study of heavy metals in ambient air by WD-XRF spectrometry using aerosol-generated filter standards, X-Ray Spectrom., 32, 129–138.

[40] Newhook R., Hirtle H., Byrne K. and Meek M. E., 2003, Releases from Copper Smelters and Refineries and Zinc Plants in Canada: Human Health Exposure and Risk Characterization, The Science of the Total Environment 301, 23–41.

[41] Var F., Narita Y. and Tanaka S., 2000, The concentration, trend and seasonal variation of metals in the atmosphere in 16 Japanese cities shown by the results of National Air Surveillance Network (NASN) from 1974 to 1996, Atmospheric Environment, 34, 2755-2770.

[42] Lau O. W., Luk S. F., 2001, Leaves of *Bauhinia blakeana* as indicators of atmospheric pollution in Hong Kong, Atmospheric Environment 35, 3113-3120.

[43] Ho K.F., Lee S.C., Chan C. K., Yu J. C., Chow J. C. and Yao X.H., 2003, Characterization of chemical species in PM2:5 and PM10 aerosols in Hong Kong, Atmospheric Environment, 37, 31–39.

[44] Ye B., Ji X., Yang H., Yao X., Chan C. K., Cadle S. H., Chan T., Mulawa P. A., 2003, Concentration and chemical composition of PM2.5 in Shanghai for a 1-year period, Atmospheric Environment, 37, 499–510.

[45] Hien P.D., Binh N.T., Truong Y., Ngo N.T., Sieu L.N., 2001, Comparative receptor modelling study of TSP, PM and PM in Ho Chi Minh City, Atmospheric Environment, 35, 2669-2678.

[46] Kumar A.V., Patil R.S., Nambi K.S.V., 2001, Source apportionment of suspended particulate matter at two traffic junctions in Mumbai, India, Atmospheric Environment, 35, 4245–4251.

[47] Maenhaut W., Ridder De D. J. A., Fernandez-Jimenez M.-T., Hooper M. A., Hooper B., Nurhayati Ms, 2002, Long-term observations of regional aerosol composition at two sites in Indonesia, Nuclear Instruments and Methods in Physics Research B, 189, 259–265.

[48] Fang G.-C., Chang C-N b, Yuh-Shen Wua, Peter Pi-Cheng Fuc,Yang C-J, Chen C-D, Chang S-C., 2002, Ambient suspended particulate matters and related chemical

species study in central Taiwan, Taichung during 1998–2001 , Atmospheric Environment, 36, 1921–1928.

[49] Fang G-C, Chang C-N, Chu C-C, Wu Y-S, Fu P. P-C, Chang S-C, Yang I-L., 2003, Fine (PM2:5), coarse (PM2:5–10), and metallic elements of suspended particulates for incense burning at Tzu Yun Yen temple in central Taiwan, Chemosphere, 51, 983–991.

[50] Fang G-C, Wu Y-S, Huang S-H, Rau J-Y, 2004, Dry deposition (downward, upward) concentration study of particulates and heavy metals during daytime, nighttime period at the traffic sampling site of Sha-Lu, Taiwan, Chemosphere, 56, 509–518.

[51] Chen S-J, Hsieh L-T, Tsai C-C, Fang G-C, 2003, Characterization of atmospheric PM10 and related chemical species in southern Taiwan during the episode days, Chemosphere, 53, 29–41.

[52] Kim K-H, Lee J-H, Jang M-S, 2002, Metals in airborne particulate matter from the first and second industrial complex area of Taejon city, Korea, Environmental Pollution, 118, 41–51.

[53] Kim, K.H., Choi, G.H., Kang, C.H., Lee, J.H., Kim, J.Y., Youn, Y.H., Lee, S.R., 2003. The chemical composition of fine and coarse particles in relation with the Asian Dust events, Atmospheric Environment 37, 753–765.

[54] European Commission Report, 2000, Ambient air pollution by As, Cd and Ni compounds, Position Paper, Office For Official Publications Of The European Communitie, Luxembourg.

[55] Nagajyoti PC., Lee KD., Sreekanth TVM., 2010, Heavy metals, occurrence and toxicity for plants: a review, Environ Chem Lett, 8, 199–216.

[56] Gebel TW, 2001, Review, Genotoxicity of arsenical compounds, Int. J. Hyg. Environ. Health 203, 249–262.

[57] International Programme on Chemical Safety (IPCS) (1992). Inorganic Arsenic Compounds other than arsine. Health and safety guide No. 70. World Health Organisation. Geneva.

[58] International Agency for Research on Cancer (IARC) - Summaries and Evaluations (1987). Arsenic and arsenic compounds. Lyon.

[59] World Health Organisation (WHO) (2001). Environmental Health Criteria Document 224. Arsenic and Arsenic Compounds. International Programme on Chemical Safety (IPCS).

[60] Floyd P., 2000, The Risks to Health and Environment by Cadmium used as a Colouring Agent or a Stabiliser in Polymers and for Metal Plating, Final report, The European Commission. UK.

[61] Valko, M., Morris, H., and Cronin, M. T. D., 2005, Metals, toxicity and oxidative stress. Curr. Med. Chem., 12, 1161–1208.

[62] Shoeters G., et al., 2006, Cadmium and children: Exposure and health effects, cadmium toxicity, Acta pediatrica, 95 Suppl. 453: 50-54

[63] Bernard A., 2008, Cadmium and its adverse effects on human health, Indian J Med Res 128, 557-564.

[64] IARC. 1993, International Agency for Research on Cancer. Cadmium and cadmium compounds. In IARC Monographs on the Evaluation of Carcinogenic Risks to Humans. Vol 58. Beryllium, Cadmium, Mercury, and exposures in the glass manufacturing industry. Lyon, 1993; 119-237.

[65] WHO (2000). Cadmium. In: Air quality guidelines for Europe, 2nd ed. Copenhagen, World Health Organization Regional Office for Europe (http://www.euro.who.int/__data/assets/pdf_file/0005/74732/E71922.pdf).

[66] EPA, 1998, Toxicological Review Of Hexavalent Chromium, U.S. Environmental Protection Agency Washington, DC

[67] ICDA, 2007, Health Safety and Environment Guidelines for Chromium, International Chromium Development Association, Paris, France

[68] Encyclopedia of occupational health and safety, 1998, 4th edition, Vol. 3, Editor; Stellman J.M, Geneva, International labour office.

[69] Langard, S. 1990. One hundred years of chromium and cancer; A review of the epidemiological evidence and selected case reports. Am. J. Ind. Med. 17, 189–215.

[70] Axelson O and Szoberg A., 1979, Cancer incidence and exposure to iron oxide dust, J Occup Med. 21(6):419-22.

[71] ATSDR, Agency for Toxic Substances and Disease Registry, ilocis, Chapter 63 - Metal: Chemical Properties and Toxicity.

[72] Sjs. Flora, 2002, Lead exposure: health effects, prevention and treatment, J. Environ. Biol. 23 (1) 25-41.

[73] Schwartz J., 1994, Low-level lead exposure and children's IQ: A meta analysis and search for a threshold, Environ. Res. 65, 42-55.

[74] Fu H and Boffetta P, 1995, Cancer and occupational exposure to inorganic lead compounds: a meta-analysis of published, Occupational and Environmental Medicine 52:73-81.

[75] Bradman A., et al., 2001, Iron deficiency associated with higher blood lead in children living in contaminated environments, Environ. Health Perspect., 109(10), 1079-1084.

[76] Centers for Disease Control and Prevention. 2006, Adult Blood Lead Epidemiology and Surveillance - United States, 2003 - 2004. Morb Mortal Wkly Rep, 55, 876-879.

[77] Committee on Environmental Health, 2005, Lead Exposure in Children: Prevention, Detection, and Management, PEDIATRICS,116(4) 1036-1046.

[78] IARC, International Agency for Research on Cancer. 2006, Working Group on the Evaluation of Carcinogenic Risks to Humans, Inorganic and organic lead compounds. IARC Monogr Eval Carcinog Risks Hum. 87: 1-468.

[79] Agency for Toxic Substances and Disease Registry, 2007, Case Studies in Environmental Medicine: Lead Toxicity, Course WB 1105.

[80] An Overview of Commodity Sectors 695, http://www.epubbud.com/read.php?g=L3A4FU3X&p=203

[81] Grimsrud T.K. and Andersen A., 2010, Evidence of carcinogenicity in humans of water-soluble nickel salts, Journal of Occupational Medicine and Toxicology, 5:7

[82] International Agency for Research on Cancer (IARC), 1990, Chromium, Nickel and Welding. Vol 49. IARC. Lyon.

[83] IUPAC, 1994, report on analysis of nickel in body fluids, the Commission on Toxicology of the International Union of Pure and Applied Chemistry.

[84] Clarkson T.W. and Magos L., 2006, The Toxicology of Mercury and Its Chemical Compounds, Critical Reviews in Toxicology, 36, 609–662.

[85] Zahir et al., 2005, Low dose mercury toxicity and human health, Environ. Toxicol. Pharmacol., 20(2), 351-360.

[86] Goyer R., 1997, Toxic And Essential Metal Interactions, Annual Review of Nutrition, Vol. 17: 37-50.

[87] Clarkson TW, 2002, The Three Modern Faces of Mercury, Environmental Health Perspectives, 110 (1), 11-23.

[88] Seixas S. et al., 2005, Accumulation of mercury in the tissues of the common octopus Octopus vulgaris (L.) in two localities on the Portuguese coast, Science of the Total Environment 340, 113– 122.

[89] McIndoe SC, Dyson PJ., 2003, Metal Carbonyls, Kirk-Othmer Encyclopedia of Chemical Technology.

[90] Armit H.W., The Toxicology Of Nickel Carbonyl, Journ of Hyg. 525-551.

[91] Raymond CS, et al. 2005Inhalational Nickel Carbonyl Poisoning in Waste Processing Workers, CHEST, 108(1), 424-429.

[92] Scott LK, Grier LR, Arnold TC, et al., 2002, Respiratory failure from inhalational nickel carbonyl exposure treated with continuous high-volume hemofiltration and disulfiram. Inhal. Toxicol. 14(11),1103-1109.

[93] Tiny, 2011, symptoms of nickel poisoning, 20-23, http://www.health-essay.com/symptoms-nickel-poisoning/

[94] Fernandez A. J., Ternero M., Barragan F. J. and Jimenez J. C., 2000, An Approach to Characterization of Sources of Urban Airborne Particles Through Heavy Metal Speciation, Chemosphere Global Change Science, 2, 123-136.

[95] Fuchtjohann L., Jakubowski N., Gladtke D., Barnowski C., Klockow D. and Broekaert, J. A. C., 2000, Determination of Soluble and Insoluble Nickel Compounds in Airborne Particulate Matter by Graphite Furnace Atomic Absorption Spectrometry and Inductively Coupled Plasma Mass Spectrometry, Fresenius Journal of Analytical Chemistry, 366 (2), 142-145.

[96] Bolt H. M., Noldes C., Blaszkewicz M., 2000, Fractionation of Nickel Species From Airborne Aerosols: Practical Improvements and Industrial Applications, Int Arch Occup Environ Health, 73, 156-162.

[97] Bhat PN, Pillai KC, 1997, Beryllium environmemntal air, water and soil, Water air and soil pollution, 95, 133-146.

[98] Canepari S., Perrino C., Astolfi M. L., Catrambone M. and perret D., 2009, Determination of Soluble Ions and Elements in Ambient air Suspended Particulate Matter: Inter-Technique Comparison of XRF, IC and ICP for Sample-by-Sample Quality Control, Talanta, 77, 1821-1829.

[99] Morishita M., Keeler G. J., Wagner J.G. and Harkema J. R., 2006, Source identification of ambient PM2.5 during summer inhalation exposure studies in Detroit, MI, Atmospheric Environment, 40, 3823–3834.

[100] Pierre F., Vallayer C., Baruthio F., Peltier A., Pale S., Rouyer J., Goutet P., Aubrege B., Lecossois C., Guillemin C., Elcabache J-M., Verelle B., Fabries J-F., 2002, Specific relationship between blood lead and air leadin the crystal industry, Int Arch Occup Environ Health, 75, 217–223.

[101] Fang G-C., Wu Y-S., Huang S-H. and Rau J-Y., 2005Review of atmospheric metallic elements in Asia during 2000–2004, Atmospheric Environment, 39, 3003–3013.

[102] Flowers M. K., Bem H. and Gorecka H., 2006, Trace Metals Concentration in Size-Fractioned Urban Air Particulate Matter in Lodz, Poland. I. Seasonal and Site Fluctuations, Polish J. Environ. Stud.,15(5), 759-767.

4

Physico-Chemical Characterisation of Aerosol Particles at Canopy Level in Urban Zone

S. Despiau

LSEET-LEPI / UMR 6017/ Université du Sud Toulon-Var
France

1. Introduction

For many years, national and/or international research programs devoted to understand the effects of aerosol particles on health (COST action 633, ..) and/or climate (ACE experiments, INDOEX, TARFOX, NAMBLEX ...) have been carried out in various regions of the world and at various scale of time and space and more and more papers are published in connection with these two effects, health (Dockery et al., 1993; Dockery and Pope, 1994; Oberdörster, 2000; Pekanen et al., 1997 ; Pope and Dockery, 1999; Pope et al., 2002; Wichmann and Peters, 2000; ...) and climate (Charlson et al., 1992; Jacobson, M.Z., 2001; Kaufman et al., 2002; Penner et al., 2004; ...).

These effects are mainly the consequence of the physico-chemical characteristics of the aerosol particles that depend first on the mode (natural, anthropogenic, primary, secondary) and place (urban, industrial, rural, marine, coastal zones ...) of production. They depend also on their transformation during transport, which is mainly depending on the meteorological conditions prevailing in the region where the experiments are carried out, the occurrence of possible and specific sources and sinks all along the pathway of the air mass. Whatever the scale considered, a global watch of these effects and their possible future changes requires the availability of adequately emission inventories (François et al., 2005) and the use of numerical models which improvement assumes accurate comparisons with experimental data (Builtjes et al., 2003, Cousin et al., 2005).

The two main objectives of the regional ESCOMPTE experimental campaign (Cros et al., 2004) which took place in the south of France in the "Fos-Berre-Marseille" region in June-July 2001, were to study the numerous photochemical pollution events that occur in that region and to produce a relevant set of data for testing and evaluating regional pollution models. Among the various data sets recorded during the campaign, the one concerning the aerosol particles resulted from the measurements realised in 7 sites representative of industrial, urban, suburban and rural areas (Cachier et al., 2005). It is well known that urban zones are a strong and complex source of aerosol particles but an "urban source" is difficult to characterize precisely. In order to try to provide data characterising better one urban zone, it was decided to realise the measurements at a height representative of a mean urban "canopy level". The measurements were then carried out on the roof of a public building localized in Marseille downtown.

The main results obtained during the campaign are presented here and are considered with respect to two main questions:

Have the particles particular physico-chemical characteristics at canopy level and are these characteristics useful for modelling purposes or for emission inventories validation ?

Are IOP situations (described hereafter) leading to a specific signature in terms of particle characteristics ?

2. Experimental campaign and methodology

The experimental zone of the ESCOMPTE program covered an area of 120x120 km centred on the "Marseille-Berre" region, including an important coastal zone (Fig. 1).

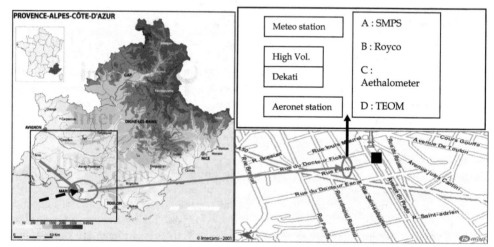

Fig. 1. Left: view of the ESCOMPTE domain (small black square), Marseille city (red ellipse); The arrows in the small square indicate the dominant wind directions. Right: Location of the measurement stations: canopy level (red ellipse) and details (above) of the sensors installed on the roof and street level (black square).

This region is characterised by the presence of numerous industrial and urban sources of primary and secondary pollutants and by specific summer meteorological conditions, mainly anticyclonic with strong solar radiation, both aspects leading to a high probability of photochemical pollution events (Cros et al., 2004). The study of these photochemical events was one of the two main objectives of the campaign and, when favourable conditions for photochemical processes were forecasted, Intensive Observation Period (IOP) was decided. Four situations: 14-15 June (IOP 1), 21-26 June [IOP 2a (21-23) and 2b (24-26)], 2-4 July (IOP 3) and 10-12 July (IOP 4), corresponded to these conditions.

2.1 Experimental set up

The building where the urban measurement station was installed is located along a one-way street canyon (Rue St Sebastien) and nearby (100 m) a two-way road (Prado avenue), both with intense traffic (Fig.1). The measurement station was installed on the roof of the building, 30 m above street level, and composed of one Scanning Mobility Particle Sizer (SMPS TSI : 3071 classifier and 3022 CPC) measuring between 14 and 720 nm, one optical counter (Royco) between 0.5 and 15 μm, two TEOMS (Rupprecht and Patachnick) with 1 and 10 μm cut-off inlets (heated at 50°C) and one aethalometer (Magee Scientific AE 14) for quasi-real time BC concentration measurements. A total filter and two cascade impactors (a 6 stages "High Vol Sierra" between 0.5 and 10 μm and 13 stages low pressure "Dekati" between 0.03 and 10 μm) were used to collect particles on filters for chemical analyse. Meteorological parameters were locally measured with an automatic station while radiative measurements were provided by a specific system located on the roof of a neighbouring building. One Aeronet station was also installed during the experimental period. Gas measurements (CO, NO_x) were taken at street level on the Prado avenue, by AIRMARAIX (now ATMOPACA), the local air pollution network.

2.2 Data acquisition

All these systems were systematically run during IOPs but also outside these specific periods, except the Dekati low pressure impactor. The SMPS and the optical counter provide number concentrations and size distributions in their specific size domains. Each one was set up to give one concentration and one distribution every 15 minutes, resulting from the average of 3 successive scans lasting 4 minutes each. Every 30 minutes, two mass concentrations (μgm^{-3}) were available with the two TEOM. The aethalometer provided one BC concentration (ngm^{-3}) every 3 minutes and five successive values were averaged to obtain one measurement every 15 minutes. In order to obtain enough material for the chemical characterization and according to their respective flow rates, the total filter, High Vol and Dekati impactors sampled 3, 6 and 10 hours, respectively. The black and organic carbon contents were determined by the 2-step thermal analysis method (Cachier et al., 1989) of the total filters, generally two, daily sampled (Cachier et al., 2005). The ionic composition was determined by chromatographic analysis (Putaud et al., 2004) of the filters impacted with the "Dekati" low pressure cascade impactor.

3. Results and discussion

3.1 PM mass concentrations

PM10 and PM1 concentration measurements started the 16/06 and 18/06, respectively, and were then monitored during the whole campaign. Over the 16/06 – 13/07 period, the average PM10 concentration (Fig. 2a) is (28 ± 5) μgm^{-3}, a value of the same order as those obtained in various European urban zones, generally comprised between 20 and 45 μgm^{-3} (Harrison et al., 2001; Perez et al., 2008; Pey et al., 2010; Putaud et al., 2004; 2010; Schwarz et al., 2008), according to the site (kerbside, downtown, suburbs), the season or to specific meteorological conditions. Averaged over 24 h only, the concentration varies between 11 μgm^{-3} the 18/06 and 61 μgm^{-3} the 06/07. During the campaign, PM10 concentration passed once only (06/07) beyond the EU standard 24-hour limit value of 50 μgm^{-3}, but reached, punctually, higher values : 131 μgm^{-3} at 16:00 and $(91,5 \pm 19,2)$ μgm^{-3} between 10:00 and 19:00, on July 6, a "non–IOP" day.

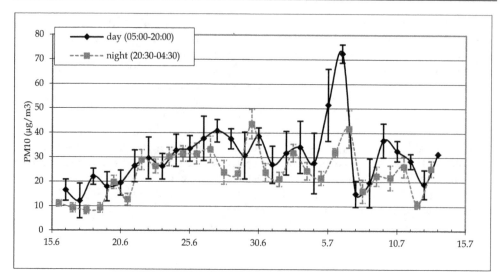

Fig. 2a. PM10 mean concentrations recorded during the campaign.

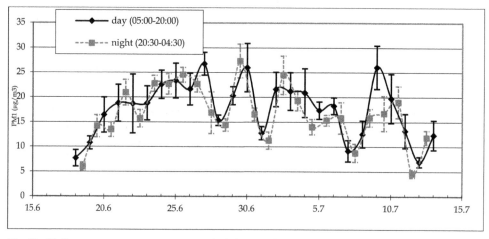

Fig. 2b. PM1 mean concentrations recorded during the campaign.

The PM1 mean concentration (fig. 2b) over the 18/06 – 13/07 period is (17.5 ± 2.1) μgm⁻³ and varies between 7 and 26 μgm⁻³ when averaged over 24 hours. These values are comparable with those obtained in Barcelona and reported by Viana et al. (2005), Pey et al. (2010) between 7 and 20 μgm⁻³, or by Gugliano et al. (2005) in Milano, around 15.3 μgm⁻³ in summer. Nevertheless, as for PM10, PM1 punctually reached higher values: 56 μgm⁻³ at 06:00 on June 22 or (35.6 ± 3.7)μgm⁻³ between 7:00 and 13:00 on June, 30. It must be noted that the higher concentrations were not recorded the same day than those of PM10 but, as for PM10, during non "IOP days". This shows that the IOP conditions do not correspond to a systematic enhancement of PM10 and PM1 concentration, and that PM10 and PM1 may evolve differently.

Analysed over 24 hours, PM10 and PM1 present both a strong increase (around 50 % to 60 % of the background values) in the morning, around 7:00 to 8:00, linked to the traffic rush. But both are also characterised by large daily variations, during and outside IOP periods. For PM10 the standard deviation represents, in average, 47% of the mean daily value and reach 67% on July, 4. The corresponding values for PM1 are 36% and 54% on June, 22. Despite the fact that PM10 includes PM1, sometimes PM10 increases while PM1 decreases and conversely (Fig. 3). In the first case this means that coarse particles (> 1µm) are mainly generated and, conversely, that fine particles (<1 µm) are generated in excess, in the second case. This simply confirms that the main PM10 and PM1 sources are different.

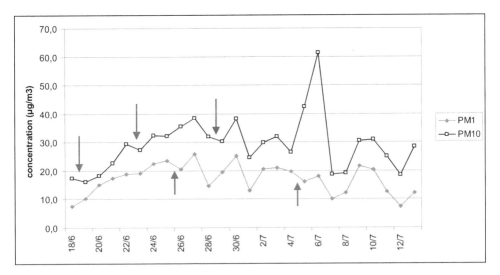

Fig. 3. Day to day comparison of PM10 and PM1 evolutions. The red arrows indicate the cases where PM1 increase while PM10 decrease, the blue arrows indicate the reverse situation.

The PM10 and PM1 concentrations depend also highly on various other factors, of which the meteorological conditions. Indeed, both wind and humidity affect differently the fine and coarse particles.

The daily comparison between the mean values of concentration and wind velocity (V), (Fig. 4.a) shows that PM10 and V are not correlated (R^2 = 0.01), that PM1 decreases slightly when V increases (R^2 = 0.41) whereas the coarse fraction (PM10 – PM1) tends to increase with V (R^2 = 0.58). This indicates that the wind velocity tends to disperse the fine particles (< 1µm) more efficiently than it generates or raises up to the canopy coarse particles, leading to the non-correlation PM10-wind velocity.

On the other hand, the PM-humidity relationship shows (Fig. 4b) that the fine particles are more humidity dependent (R^2 = 0.54) than the coarse (R^2 = 0.10 only), indicating a less hygroscopic character of the larger particles.

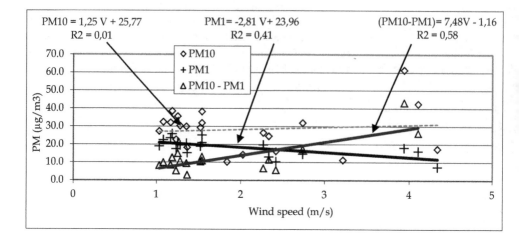

Fig. 4a. PM10, PM1 and (PM10-PM1) versus wind speed.

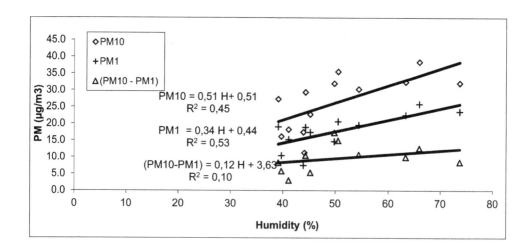

Fig. 4b. PM10, PM1 and (PM10-PM1) versus humidity.

The comparison between the fine and coarse fractions is generally made through the study of PM2.5/PM10 ratio (Harrison et al., 2001; Keywood et al., 1999; Putaud et al, 2010; Querol et al., 2004; Van Dingenen et al., 2004) and more rarely through the PM1/PM10 ratio. In our case, the PM1/PM10 ratio varies mainly between 0.5 and 0.7 leading to an average value over the whole campaign of (0.60 ± 0.12). These values are in accordance with those of Gugliano et al. (2005) in Milano, around 0.60, but slightly greater than those of Keywood et al. (1999) for six cities in Australia (0.50 ± 0.08) or those by Viana et al. (2005), between 0.33 and 0.54 and Perez et al. (2008) between 0.42 and 0.48, both in Barcelona. However, this ratio reached during the campaign higher (0.85 the 20/06) or lower values, (0.3 the 06/07) and increased slightly (0.69 ± 0.06) during IOPs 2 and 3, indicating a tendency for IOP periods to produce more fine particles.

By comparison, the average PM2.5/PM10 ratio calculated by Putaud et al. (2004) from measurements in various European regions including rural and urban sites is (0.73 ± 0.02), but it is précised (Putaud et al, 2010) that this ratio may vary between 0.4 and 0.9. These large variations make the comparisons difficult but our PM1/PM10 values are not very different of the PM2.5/PM10 ratio, showing that the (PM2.5-PM1) may be estimated around 10% only of PM10 in our situation. This estimation is comparable with the Pey et al (2009) value (15% in urban background in Barcelona) and Querol et al. (2004) values who found that the PM2.5-1 contributed between 9 and 21 % to PM10 in their study in different locations of Spain.

Nevertheless, in order to better compare the relative importance of the fine and coarse modes, we calculated the ratio PM1 / (PM10 - PM1) for each of the 24 days with available data.

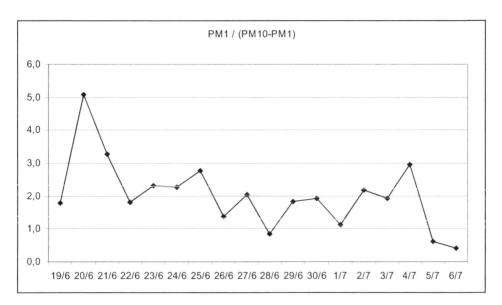

Fig. 5. Ratio PM1/(PM10-PM1) calculated for each day of the campaign.

The results plotted in Fig. 5 indicate clearly that, in our situation, the fine mode dominates largely, in mass, the coarse mode. Averaged over the whole campaign the ratio is (1.9 ± 0.8) but it may reach in some cases higher values (3.3 the 21/06 or 5.1 the 20/06), while it is < 1 for only three days. The fact that the measurements are realised at the canopy level may partially explain this result since the coarse particles are more efficiently deposited and/or less efficiently raised up than the fine particles, in particular during the night and periods of weak winds.

3.2 Chemical composition

3.2.1 Carbon compounds

Elemental or black carbon (EC or BC) measurements were everyday available through the use of the aethalometer. In parallel with the aethalometer measurements, EC and total carbon (TC) were also daily available through total filters analysed by the thermal method (Cachier et al., 1989). We dispose also of one set of "Dekati filters" sampled during the 23-24 June, in the middle of IOP2, allowing in that case the study of EC and TC mass distribution. Organic carbon (OC) was deduced from the previous values (OC = TC - EC).

Over the whole campaign, the average EC concentrations measured by the aethalometer and deduced from the filter analyse are comparable, (2.93 ± 0.73) and (2.75 ± 0.92) μgm^{-3}, respectively. The TC concentration obtained from the filter analyse is (9.11 ± 2.38) μgm^{-3}, the corresponding OC value is thus evaluated at (6.37 ± 1.95) μgm^{-3}. As for PM1 or PM10, to compare our TC and/or EC measurement results with the large number of those realised in various places in the world is quite difficult because of the specific conditions (meteorological, season, site, duration) that characterise each measurement.

Our TC values are nevertheless comprised in the wide range of those obtained in similar sites (urban or urban background) and summer season, which vary between 4 to 5 μgm^{-3} in Aveiro (Castro et al., 1999), Helsinki (Viidanoja et al., 2002), Belfast (Jones and Harisson, 2005) or in Oporto (Duarte et al., 2008) and higher values around 19 μgm^{-3} in Seoul (Kim et al., 1999) or Hong Kong (Ho et al., 2002). The EC/TC ratio is equal to (0.31 ± 0.08). This ratio is higher than the one (0.17) reported by Pey et al. (2009) for regional background sites in the west coast of Spain but equivalent to those reported by Viana et al. (2007) concerning measurements in Amsterdam and Barcelona, or by Jones and Harrison (2005) in the centre of Belfast and of the same order than those generally found in measurements carried out in similar urban zones, comprised between 0.22 and 0.38, and reported in Ruellan and Cachier, (2001), Yttri et al. (2007) or Harrison and Yin (2008). For the IOP days only, the results are slightly different. The EC, TC and OC concentrations are respectively (2.72 ± 0.94) μgm^{-3}, (10.37 ± 2.59) μgm^{-3} and (7.65 ± 1.97) μgm^{-3}, while the EC/TC value is 0.26 ± 0.07. The tendency for IOP days is thus an increase of the OC concentrations and, consequently, a decrease of the EC/TC ratio. This tendency is logical since the IOP days are favourable to photochemical processes that favour secondary organic aerosol formation and thus the increase of the OC concentration that is partly composed of secondary organic aerosols.

3.2.2 Water Soluble Compounds (WSC)

Seven sets (6 days and 1 night) of 13 filters sampled with the Dekati impactor during IOP1 (14 and 15/06), IOP2 (22, 25d, 25n, and 26/06) and IOP3 (02/07) periods, were analysed by ionic chromatography. The amount of the different water soluble compounds in PM10 and PM1 was obtained by summing, for each set of 13 filters, the concentrations measured on each one. Averaged over the 7 samples, nss sulphate (nssSO$_4$) is the most abundant (5.7 ± 2.3 µgm^{-3}), followed by sea salt (3.81 ± 1.19 µgm^{-3}), ammonium and nitrates in equivalent amount, (2.9 ± 0.4 µgm^{-3}) and (2.8 ± 1.1 µgm^{-3}), respectively, then by the main crustal elements [nss(Ca + Mg + K) ≈ 1.4 µgm^{-3}]. These values are of the same order than those obtained in Barcelona (Viana et al., 2005; Perez et al., 2008) or in other coastal sites in Spain (Querol et al., 2004). These average values hide however the large variations observed on some occasions. The 25 June (IOP 2b) the nssSO$_4$ and NH$_4$ concentrations reach respectively 8.7 and 4.22 µgm^{-3}, while they are of 1.91 and 1.09 µgm^{-3} only the 22 June (IOP 2a). Inversely, the sea salt concentration is maximum (5.5 µgm^{-3}) the 22 June and of 3.8 µgm^{-3} only the 25 June. These variations result mainly from the very different meteorological conditions of these two days. The 22, the average humidity during the sampling time is 41.8% only while it is 78.5% the 25. The wind velocity is 5.5 ms^{-1} in average between 11:30 and 17:30 the 22, and of 2.7 ms^{-1} only the 25 for the same period. Moreover, the mean wind direction the 22 corresponds to a situation of sea breeze leading to an enhanced production of coarse marine aerosol while the direction is more continental the 25, coming mainly from the industrial Fos-Berre zone.

3.2.3 Mass chemical distribution

The use of the cascade impactor allows the study of the mass distribution of the water soluble compounds for the 7 samples available and of the carbon compounds for the 23-24 July set of filters. 80% of EC and 69% of OC are found on the fine mode (<1µm).

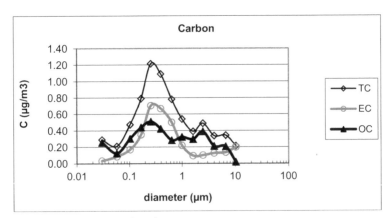

Fig. 6. TC, EC and OC mass size distributions.

Our results (Fig. 6) show that the EC distribution is monomodal and centred on the accumulation mode on particles of diameter comprised between 0.25 and 0.4 µm, while the OC distribution tends to be bimodal with two relative maximum at 0.25 and 2.5 µm, respectively. Nevertheless, that distribution must be considered as indicative because they are deduced from the analysis of one set of filters only.

For the water soluble compounds, the analyse shows that 100 % of ammonium and 94 % of nss-sulphate are found in the fine mode and for the major part on particles of diameter 0.3-0.4 μm. Sea salt and crustals are classically found on the coarse mode, both at 80 %, on particles of diameter 1 and 5 μm, while nitrate is spread over the whole range but mainly, at 64 %, on the coarse mode (Fig.7). Concerning NO_3 it must be noted the fine particle amount is mainly due to the distribution obtained during the night (25/06) and that its distribution in the coarse mode tends to be bimodal with two maximum around 2.5 and 7 μm.

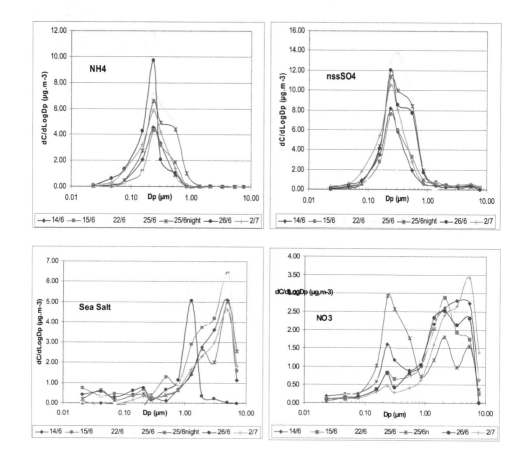

Fig. 7. Mass size distribution of 4 WSC corresponding to the seven sets of filters.

In Fig. 8 are reported the relative percentages of the compounds analysed for PM10, PM1 and PM1-10.

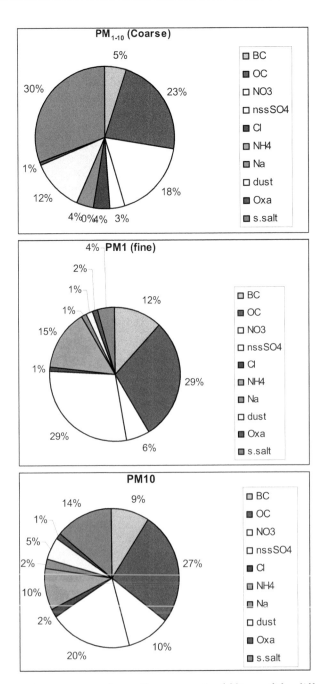

Fig. 8. relative percentages, averaged over the seven sets of filters, of the different compounds in PM10, PM1 (Fine) and PM(1-10) (coarse).

It can be seen that for PM10 (36%) as for PM1 (41%), TC (OC + EC) is by far the major contributor, followed by nssSO$_4$, 20% and 29%, respectively. For PM1-10, sea salt becomes the major contributor (30%), followed by TC (28%). Compound by compound, the four more important are OC (27%), nssSO4 (20%), seasalt (14%) and NH$_4$+ and NO$_3$ (10%) for PM10, OC (29%), nssSO$_4$ (29%), NH$_4$ (15%) and EC (12%) for PM1, seasalt,(30%), OC (23%), NO$_3$ (18%) and crustals (14%) for PM1-10. The comparison between the three distributions shows that, for PM1, the relative parts of carbon (EC and OC), nss SO$_4$ and NH$_4$ increase, while NO$_3$ and sea salt decrease, confirming simply that the first compounds are mainly distributed on the fine particles and conversely for NO$_3$ and sea salt.

3.2.4 Correlation between WSCs

We calculated (Table 1) the coefficient of correlation between different WSCs for each set of filters (the 13 filters are considered together) and for each day. When R^2 is very high (> 0.8), this means that the two compounds are found on the same size category, fine or coarse, and more precisely on the same narrow size range.

R^2	14/06	15/06	22/06	25/06D	25/06N	26/06	02/07
NH$_4$ - nssSO$_4$	0.99	0.98	0.98	0.98	0.99	0.44	0.98
NH$_4$ - NO$_3$	0.05	0.07	0.32	---	0.56	0.18	0.30
Na - NO$_3$	0.84	0.86	0.45	0.48	---	0.34	0.73
Na - Cl	0.58	0.55	O.97	0.35	0.35	0.06	0.40

Table 1. Coefficient of correlation between different WSCs.

It is clearly the case for NH$_4$ and nss SO$_4$ that are, except the 26/06, both localised on particles of diameter comprised between 0.3 and O.5 µm and classically linked as (NH$_4$)$_2$SO$_4$. It is also the case for Na and NO$_3$ the 14 and 15 June, associated as NaNO$_3$ in coarse particles or Na and Cl (NaCl) the 22 June, also on the coarse mode. On the contrary, when R^2 is weak, the two compounds are distributed differently (coarse and fine). However, even when R^2 calculated over the whole size range is weak, it is possible that the corresponding compounds be associated only in the fine or in the coarse mode. For example, R^2 calculated for NH$_4$ and NO$_3$ over the fine mode only, reach values > 0.90 the 14/06, 15/06 or 25/06 of June, indicating the presence of ammonium nitrate in the fine mode. The 26/06 is characterised by a weak ammonium-nss-sulphate correlation (R^2 = 0.44) and no ammonium-nitrate correlation. According to wind direction and back-trajectories, the air masses on that day were exclusively of marine origin that could explain the weak correlations, as it was previously noticed in the same region and under similar conditions by Sellegri et al. (2001).

3.3 Particle concentrations and size distributions

During the campaign, the SMPS system was set up to provide one average Size Distribution (SD) every 15 minutes, derived from 3 SD lasting 4 minutes each. From these SD, other parameters were deduced (total concentrations, day average SD,...), according to the analysis conducted. Total dN or dN/dLogD concentrations were obtained by summing the concentrations measured for each size range. In Fig. 9 are plotted the day and night average of the dN concentrations obtained at the canopy level for 17 days and 10 nights during the campaign.

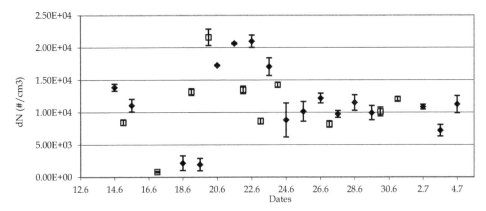

Fig. 9. Plot of the day (black stars) and night dN average concentrations (with standard deviation) recorded during the campaign.

These concentrations vary mainly between $0.75.10^4$ and $1.4.10^4$ cm^{-3}, except in two occasions. The 18/6 and 19/6 the concentrations are lower because these two days are characterised by strong winds [(8 ± 1) ms^{-1} in average against ≈ 2 ms^{-1} for the other days] that disperse very efficiently the fine particles, as already noted for the PM1 concentration. The 21/6 and 22/6, which are the first days of the IOP2, the concentrations are on the contrary higher. Averaged over the whole campaign the concentration is $(1.3 ± 0.5)10^4$ cm^{-3} including the 4 specific situations mentioned before or $(1.3 ± 0.2).10^4$ cm^{-3}, without. This value, that may be considered as the background value at the canopy level for the corresponding period and site, is of the same order than those measured in Prague ($1.18\ 10^4$ cm^{-3}) by Salma et al. (2011), in Augsburg ($1.22\ 10^4$ cm^{-3}) by Birmili et al. (2010) or in other urban sites (London, Leipzig or Milano), comprised between $0.7.10^4$ and $2.4.10^4$ cm^{-3}) and reported in Putaud et al., (2004, 2010). Our value is however largely lower than those measured at kerbside, which vary between $4.0.10^4$ and $6.0.10^4$ cm^{-3}, Putaud et al., (2004, 2010) or by Yue et al., (2010), 2.910^4 cm^{-3} in average in Ghangzoua, during summer pollution episodes. This simply shows that, in urban zone and in summer season, the particles produced by the traffic at street level are partially dispersed before reaching the canopy and that measurements made at canopy level are likely more representative, as input values for regional models, than results obtained at street level. It must also be noted that IOP days, as it was already the case for the PM measurements, do not correspond systematically to the higher concentration values. Indeed, the concentrations obtained during the IOP1, 2b and 3 are of the same order than those of "non-IOP" days.

If these daily average values give a quite good indication of what is expected at the canopy level of a great coastal urban area in summer and may be considered as useful input values for modelling purpose at local and regional scale, or for comparison with particle emission inventories in urban zone, they hide nevertheless the large variations that occur generally during a day.

In Fig. 10 are plotted the background estimated value $((1.3 \pm 0.5)10^4$ cm^{-3} - dot point line) and the variations recorded, as an example, the 21 June. Two parts, marked by ellipse on the figure, are typical and detected every day. The part 1 corresponds to the morning traffic rush characterised by a strong increase in concentration detected at roof level between 7:00 and 8:00. The part 2 corresponds to the period of lowest concentration due to the conjugate effect of wind speed increase (from ≈ 2 ms^{-1} in the morning to ≈ 4 -5 ms^{-1} in the afternoon), often linked to the afternoon sea breeze regime, and the vertical development of the boundary layer that disperse very efficiently the particles.

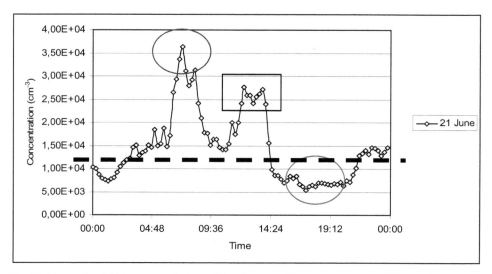

Fig. 10. Example of 24 hours evolution of the fine particles concentration. Ellipses blue and pink mark the maximum and minimum daily observed. The black rectangle mark a specific case of secondary aerosol production.

The peak around 12:00, discussed in detail in Despiau and Croci (2007), is more specific. It is independent of any traffic rush but corresponds to the signature of secondary aerosol production process that occurs in special conditions of wind direction, humidity and radiation. For the six days where that specific peak was detected (hereafter "SPD", Secondary Production Day), the dN average concentration is $(1.12 \pm 0.7)10^4$ cm^{-3} while it is of $(1.05\pm 0.4)10^4$ cm^{-3} for "normal" days (hereafter "ND"). These values remain of the same order than those indicated in the previous paragraph. This means that the diurnal average values remain practically constant, even when a specific process occur, and then do not allow to consider short time variations. Indeed, during the morning rush, the mean of the concentration maximum is around 3 times the previous value: $(3.3 \pm 1.7)10^4$ cm^{-3} while during the afternoon the concentration is only one half of the mean diurnal value.

The dN or dN/dLogD shape of the particle size distributions gives also useful indications about the characteristics of the particles. In Fig. 11 are reported the averaged size distributions obtained during the night (9 nights between 20:00 and 06:45), during the 6 days with secondary production (SPD) and during the "normal" 10 days (ND). The normal and secondary production days differ mainly by their diameter mode, (93.0 ± 15.5) nm for ND, (40 ± 4) nm only for SPD. On the contrary, the mode concentrations are of the same order (1.24 ± 0.5) 10^4 cm^{-3} for ND, and (1.46 ± 0.94) 10^4 cm^{-3}, for SPD. In-between (ND) and (SPD) is the average night distribution with a mode diameter about 55 nm and a concentration around 1.3 10^4 cm^{-3}.

The size distributions corresponding to the specific periods illustrated by the Fig. 10 are also characteristic and largely different from the mean size distributions of Fig. 11.

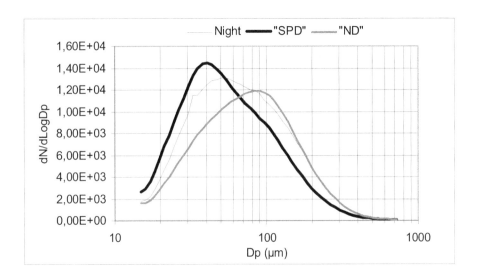

Fig. 11. Plot of the mean size distribution corresponding to "normal" days (ND), "secondary production day" (SPD) and night.

Since they have been presented in details in Despiau and Croci (2007) we just summarize here their main characteristics. The size distribution of the morning traffic rush are characterized by an increase (factor 3 to 4 with respect to the background values) in concentration of the whole size range of fine particles leading to a monomodal SD with a diameter mode around 100 nm. For those corresponding to the secondary production process, only the concentration of the particles smaller than 70 nm increases (factor 4 in 30 to 45 minutes) leading to a peak around 35 nm in average, Fig. 12.

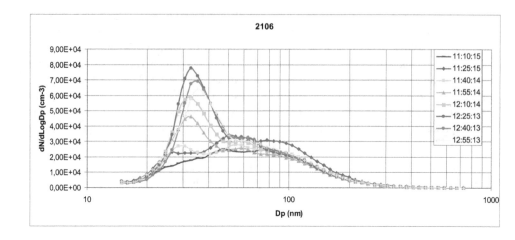

Fig. 12. Evolution of the size distribution in one case (21/06) of secondary aerosol production.

3.4 Discussion with regard to an "urban signature"

All the previous results come from measurements realised at the canopy level. They are obviously a mixing of local urban sources and conditions, and of regional transport from the surroundings, that may be, in our situation, as different as marine or anthropogenic origin. It is then difficult to separate precisely the relative influence of the different sources but we estimate that the results obtained reflect quite accurately the main physico-chemical characteristics of the particles at that level and under the specific conditions described.

So, considering the two main questions asked in the introduction, the results obtained show clearly that there is not a specific physical signature corresponding to the local IOP conditions. Nor the PM10 and PM1 mean concentrations are significantly different (larger or smaller for example) for IOP or non-IOP days. The mean number concentrations or size distributions are not either different. We detected a normal tendency for enhanced concentration in OC during IOP days but without conduct to mean concentrations significantly different (7.67 ± 1.97 μgm^{-3} against $6.37 \pm 1.95 \mu gm^{-3}$). We cannot conclude about the WSC chemical composition since we measured only during IOP days but it seems unlikely that the chemical composition be very different, except perhaps for compounds that would mainly result, as OC, from strong photochemical processes.

It is nevertheless possible to describe the main physico-chemical characteristics of the aerosol particles at the canopy level of a great Mediterranean conurbation, in summer. These local conditions may appear restrictive but we believe that similar conditions may be found at least in the Mediterranean area where various great cities are coastal cities (Barcelona,

Valencia, Genoa, Roma, Athens or north African cities, …) and probably in other coastal regions in the world where conditions leading to photochemical processes would be fulfilled. For modelling at large scale and over long periods and probably to compare with emission inventories at urban zone scale, the order of magnitude of the mean values obtained should be useful. They may be summarized in terms of PM10 and PM1 concentrations around 28 and 17 μgm^{-3}, respectively, number concentration around 1.3 10^4 cm^{-3} and a unimodal size distribution with a diameter mode around 100 nm. Always in terms of order of magnitude, the concentrations of the main elements may be estimated at 10 μgm^{-3} for carbon compounds (7 μgm^{-3} for OC and 3μgm^{-3} for EC), 6 μgm^{-3} for nss SO_4, 4 μgm^{-3} for sea salt and 3 μgm^{-3} for ammonium and nitrate. Ammonium and nssSO4 are distributed over the fine particles, sea salt mainly over coarse particles while nitrate and carbon compounds are spread over the whole size range.

However, if more precise information are needed for modelling at local scale and/or short time period for example, or to verify estimations of inventory emissions at short time scale, the previous average data become unreliable. It is then necessary to take into account specific characteristics like those linked to the morning traffic rush, the afternoon decrease, the secondary aerosol processes or very particular meteorological conditions. All those particular situations last about 2 to 3 hours and are characterized by very different values of PM or dN concentrations, shape of size distribution or mode diameter. Inversely, a significant modification of the chemical composition over the same time scale is quite unlikely since this would suppose a change in the sources that supply the particles and because the chemical modification of the particles is a process much more slow than dN or PM concentration variations. During the morning traffic rush or the SPD events, the dN and PM concentrations will increase strongly while the corresponding sources do not change.

Nevertheless, the meteorological conditions play a crucial role that must be accounted for, especially at local scale. The wind velocity and the vertical development of the boundary layer will influence the afternoon decrease. The wind velocity and direction will contribute to the transport over the city of particles from different origin (marine or industrial in our situation). The amount of radiation will influence the production of secondary aerosols while the humidity will influence the mass concentration or the chemical composition. So, except for the morning traffic rush that is well characterized whatever the meteorological conditions, most of the other situations must be envisaged only after considering the main meteorological conditions: wind velocity and direction, radiation and humidity.

4. Conclusion

The measurements carried out in urban zone during the ESCOMPTE campaign allowed us to study, at roof level and summer season, the physico-chemical characteristics of aerosol particles.

Most of the number or mass concentrations, chemical composition or size distributions results obtained during this campaign, are in agreement with those recorded in large urban zones and reviewed in the papers by Putaud et al. (2004, 2010). Nevertheless, the PM1 and

PM10 mass concentrations are characterized by strong daily variations, they may sometimes evolve in opposite sense and the fine mode (D_p < 1μm) is largely dominant in mass compared to the coarse mode. The fine particles are more dispersed by the wind and are more humidity dependent than those of the coarse mode.

During the campaign, the OC concentration was the higher, followed by nss sulphate, sea salt, ammonium nitrates and EC in equivalent concentration and finally by the main crustal elements. EC, nss sulphate and ammonium are found on the fine mode, sea salt and crustal elements on the coarse mode while nitrates and OC are spread over the two modes. The OC concentration tends to increase in the fine mode during the periods of strong photochemistry conditions that produce secondary aerosols.

The background number concentration at the canopy level has been estimated around (1.3 ± 0.5) 10^4 cm^{-3}. This value is significantly lower than concentrations measured at kerbsite but do not vary significantly according to the environmental conditions, except when the day average wind velocity go beyond 4 to 5 ms^{-1}.

That background value hide however the large variations observed daily and characterized by a strong increase in the morning, linked to the morning traffic rush, and by a minimum in the afternoon explained by the double dispersion effect of wind and boundary layer dynamics.

On some occasions a peak in concentration resulting from secondary aerosol production has been detected and lead to a size distribution with a mode around 40 nm while on "normal" days the mode is around 100 nm.

The results obtained also show that it does not exist, at canopy level in urban zone, a typical or characteristic "aerosol particle signature" corresponding to the specific conditions leading to photochemical events. Nevertheless, it is possible to describe a set of average characteristics that might be useful for modelling purposes at large spatial or time scale and to be compared with emission inventories.

For similar considerations at shorter time or spatial scales, it becomes necessary to take into account specific characteristics like the increase in concentration linked to the morning traffic rush, the afternoon decrease linked to the wind velocity and the development of the boundary layer, both every day detected, or, in some particular circumstances, the effect of secondary aerosol production.

For those situations, except for the morning rush, the meteorological conditions influence is obvious. The wind velocity determines the dispersion process, thus the mass and number distributions measured at roof level. The chemical composition may be modified according to wind direction. The relative humidity modify the mass distribution, while the radiation level may contribute to photochemical phenomenon and secondary aerosol production.

5. Acknowledgements

This work has been funded by two French national research programmes : PNCA and PRIMEQUAL-PREDIT and by the ADEME agency. The author would thanks very specially H. Cachier and J.P. Putaud for their help in chemical analyse.

6. References

Birmili, W., Heinke, K., Pitz, M., Matschullat, J., Wiedensohler, A., Cyrys, J., Wichmann, H.-E., and Peters, A., (2010). Particle number size distributions in urban air before and after volatilisation, Atmos. Chem. Phys., 10, 4643-4660.

Builtjes, P.J.H., Borrego, C., Carvalho, A.C., Ebel, A., Memmesheimer, M., Feichter, H., Münzeberg, A., Schaller, E. & Zlatev, Z. (2003). Global and regional atmospheric modelling, overview of subproject GLOREAM. In *Towards Cleaner Air for Europe – Science, Tools and Applications*, P.M. Midgley and M. Reuther ed., 139-164.

Cachier, H., Aulagner, F., Sarda, R., Gautier, F., Masclet, P., Besombes, J.L., Marchand, N., Despiau, S., Croci, D., Mallet, M., Laj, P., Marinoni, A., Deveau, P.A., Roger, J.C., Putaud, J.P., Van Dingenen, R., Dell'acqua, A. Viidanoja, J., Martins-Dos santos, S., Liousse, C., Cousin, F., Rosset, R., Gardat, E., & Galy-Lacaux, C. (2005). Aerosol studies during the ESCOMPTE Experiment : an Overview. *Atmospheric Research*, 74, 547-563.

Cachier, H., Brémond, M.P. & Buat-Menard, P. (1989). Determination of atmospheric soot carbon with a simple thermal method. *Tellus* 41B, 379-390.

Castro, L.M., Pio, C.A., Harrison, R.M., & Smith, D.J.T.(1999). Carbonaceous aerosol in urban and rural European atmospheres: estimation of secondary organic carbon concentrations. *Atmospheric Environment* 33, 2771-2781.

Charlson, R.J., Schwartz, S.E., Hales, J.M., Ces, R.D., Coakley Jr., J.A., Hansen, J.E., & Hoffman, D.J. (1992). Climate forcing by anthropogenic aerosols. *Science*, 255, 423-430.

Cousin, F., Liousse, C., Cachier, H., Bessagnet, B., Guillaume, B. & Rosset, R. (2005). Aerosol modelling and validation during ESCOMPTE 2001. *Atmospheric Environment* , 39, 1538-1550.

Cros, B., Durand, P., Cachier H., Dobrinski, P., Frejafon, E., Kottmeier, Perros, P.E., Peuch, V.H., Ponche, J. L., Robin, D., Saïd, F., Toupance, G. & Wortham, H. (2004). An overview of the ESCOMPTE campaign. *Atmospheric research* , 69, 241-279.

Despiau, S. & Croci, D. (2007). Concentrations and size distributions of fine aerosol particles measured at roof level in urban zone. *Journal of Geophysical Research*, vol. 112, D09212, doi:10.1029/2006JD007228.

Dockery, D.W., Pope, A, Xu, X., Spengler, J.D., Ware, J.H., Fay, M.E., Ferris, B.G. & Speizer, F.E. (1993). An association between air pollution and mortality in six US cities. *New England Journal of Medicine* , 329, 1753-1759.

Dockery, D.W. & Pope, C.A. (1994). Acute respiratory effects of particulate air pollution. *Annual revue of Public Health*, 15, 107-132.

Duarte, R.M.B.O., Mieiro, C. L., Penetra, A., Pio, C. A. & Duarte, A. C. (2008). Carbonaceous materials in size-segregated atmospheric aerosols from urban and coastal-rural areas at the Western European Coast. *Atmospheric Research*, 90, 2-4, 253-263

François, S., Fayet, S., E. Grondin & Ponche, J.L. (2005). The establishment of the atmospheric emission inventories of the ESCOMPTE program, *Atmospheric Research*, 74, 1-4, 5-35.

Gugliano, M., Lonati, G., Butelli, P., Romele, L., Tardivo, R & Grosso, M. (2005). Fine particulates (PM 2.5-PM1) at urban sites with different traffic exposure, *Atmospheric Environment*, 39, 2421-2431.

Harrison, R.M., Yin, J., Mark, D., Stedman, J., Appleby, R.S., Booker, J. & Moorcroft, S. (200)1. Studies of the coarse particle (2.5-10 μm) component in urban atmospheres. *Atmospheric Environment*, 35, 3667-3679.

Harrison, R.M. & Yin, J. (2008). Sources and processes affecting carbonaceous aerosol in central England. *Atmospheric Environment*, 42, 7, 1413-1423.

Ho, K.F., Lee, S.C., Yu, J.C., Zou, S.C. & Fung, K. (2002). Carbonaceous characteristics of atmospheric particulate matter in Hong Kong. *The Science of the Total Environment*, 300, 59-67.

Jacobson, M.Z. (2001). Strong radiative heating due to the mixing state of black carbon in atmospheric aerosols. *Nature*, 409, 695-697.

Jones, A.M. & Harrison, R.M. (2005). Interpretation of particulate elemental and organic carbon concentrations at rural, urban and kerbside sites. *Atmospheric Environment* 39, 7114-7126.

Kaufmann, Y.J., Tanré, D. & Boucher, O. (2002). A satellite view of aerosols in the climate System. *Nature*, 419, 215-223

Keywood, M.D., Ayers, G.P., Gras, J.L., Gillett, R.W. & Cohen, D.D. (1999). Relationships between size segregated mass concentration data and ultrafine particle number concentrations in urban areas. *Atmospheric Environment*, 33, 2907-2913.

Kim, Y.P., Moon, K.C., Lee, J.H.& Baik, N.J. (1999). Concentrations of carbonaceous species in particles at Seoul and Cheju in Korea. *Atmospheric Environment* , 33, 2751-2758.

Oberdörster, G. (2000). Toxicology of ultrafine particles : in vivo studies. *Philosophical Transactions of the Royal Society of London* , A358, 2719-2740.

Penner, J.E., Dong, X., & Chen, Y. (2004). Observational evidence of a change in radiative forcing due to indirect aerosol effect. *Nature*, 427, 231-234.

Perez, N., Pey, J., Querol, X., Alastuey, A., Lopez, J.M. & Viana, M. (2008). Partitioning of major and trace components in PM_{10}- $PM_{2.5}$-PM_1 at an urban site in Southtern Europe. *Atmospheric Environment*, 42, 8, 1677-1691.

Pey, J., Pérez, N., Castillo, S., Viana, M., Moreno, T., Pandolfi, M., López-Sebastián, J.M., Alastuey, A. & Querol, X. (2009). Geochemistry of regional background aerosols in the Western Mediterranean, *Atmospheric research*, 94, 3, 422-435.

Pey,J., Alastuey, A., Querol, X., Pérez, N. & Cusack, M. (2010). Monitoring of sources and atmospheric processes controlling air quality in an urban Mediteranean environment. *Atmospheric Environment*, 44, 2, 285-299

Pope III, C.A. & Dockery, D.W. (1999). Epidemiology of particle effects. *Air Pollution and Health* 31, 673-705.

Pope, C.A., Burnett, R.T., Thun, M.J., Calle, E.E., Krewski, D., Ito, K. & Thurston, G.D. (2002). Lung cancer, cardiopulmonary mortality, and long-term exposure to fine particulate air pollution. *Journal of the American Medical Association*, 287, 1132-1141.

Putaud, J.P. et al. (2004). A european aerosol phenomenology –2: chemical characteristics of particulate matter at kerbsite, urban, rural and background sites in Europe. *Atmospheric Environment* 38, 2579-2595.

Putaud et al. (2010). A European aerosol phenomenology-3 : Physical and chemical characteristics of particulate mater from 60 rural, urban and kerbside sites across Europe. Atmospheric Environment, 44, 10, 1308-1320.

Querol, X., Alastuey, A., Viana, M., Rodriguez, M., Artinano, S., Salvador, B., Garciado Santos, P., Fernandez Patier, S., Ruiz, R., de la Rosa, C.R., Sanchez de la Campaa, J., Menendez, A.,M. & Gil, J.I. (2004). Speciation and origin of PM10 and PM2.5 in Spain. *Aerosol Science,* 35, 1151-1172.

Ruellan, S., H. & Cachier (2001). Characterisation of fresh particulate vehicular exhausts near a Paris high flow road. *Atmospheric Environment* 35, 453-468.

Salma, I., Borsós, T., Weidinger, T., Aalto, P., Hussein, T., Dal Maso, M., & Kulmala, M. (2011). Production, growth and properties of ultrafine atmospheric aerosol particles in an urban environment, *Atmos. Chem. Phys.*, 11, 1339-1353

Schwarz, J., Chi, X., Maenhaut, W., Civiš, M., Hovorka, J. & Smolík, J. (2008). Elemental and organic carbon in atmospheric aerosols at downtown and suburban sites in Prague. *Atmospheric Research*, 90, 2-4, 287-302

Sellegri K., Gourdeau, J., Putaud J.P. & Despiau, S. (2001). Chemical composition of marine aerosol in a Mediterranean coastal zone during the FETCH experiment, *Journal of Geophysical Research*, 106, D11, 12023-12037.

Van Dingenen R. et al. (2004). A European aerosol phenomenology-1: physical characteristics of particulate matter at kerbside, urban, rural and background sites in Europe. *Atmospheric Environment* 38, 2561-2577.

Viana, M., Pérez, C., Querol, X., Alastuey, A., Nickovic, S. & Baldasano, J.M. (2005). Spatial and temporal variability of PM levels and composition in a complex summer atmospheric scenario in Barcelona (NE Spain). *Atmospheric Environment* 39, 5343-5361.

Viana, M., Maenhaut, W., ten Brink, H.M., Chi, X., Weijers, E., Querol, X., Alastuey, A., Mikuska, P., & Vecera, Z. (2007). Comparative analysis of organic and elemental carbon concentrations in carbonaceous aerosols in three European cities. *Atmospheric Environment*, 41, 28, 5972-5983.

Viidanoja, J., Sillanpää, M., Laakia, J., Kerminen, V.M., Hillamo, R., Aarnio, P. & Koskentalo, T. (2002). Organic and black carbon in PM2.5 and PM10: 1 year of data from an urban site in Helsinki, Finland, *Atmospheric Environment*, 36, 3183-3193.

Wichmann, H.E. & Peters, A. (2000). Epidemiological evidence of the effects of ultrafine particle exposure. *Philosophical Transactions of the Royal Society of Londo,n* A358, 2751-2769.

Yttri K. E., Aas, W., Bjerke, A., Cape, J.N., Cavalli, F., Ceburnis, D., Dye, C., Emblico, L., Facchini, M.C., Forster, C., Hanssen, J.E., Hansson, H.C., Jennings, S.C., Maenhaut, W., Putaud, J.P. & Torseth, K. (2007). Elemental and organic carbon in PM_{10}: one year measurement campaign within the European Monitoring and Evaluation Programme EMEP. *ACP*, 7, 22, 5711-5725.

Yue, D. L., Hu, M., Wu, Z. J., Guo, S., Wen, M. T., Nowak, A., Wehner, B., Wiedensohler, A., Takegawa, N., Kondo, Y., Wang, X. S., Li, Y. P., Zeng, L. M. & Zhang, Y. H. (2010). Variation of particle Number size distributions and chemical compositions at the urban and downwind regional sites in the Pearl River Delta during summertime pollution episodes. *Atmos. Chem. Phys.*, 10, 9431-9439, doi:10.5194/acp-10-9431-2010.

5

Air Quality Study, Comparison Between the Proposed and Actual Scenarios of Generator Sets in Havana, by Using CALPUFF Model

Yasser Antonio Fonseca Rodríguez[1], Leonor Turtós Carbonell[2],
Elieza Meneses Ruiz[1], Gil Capote Mastrapa[1] and José de Jesús Rivero Oliva[2]
[1]*CUBAENERGIA*
[2]*Universidade Federal do Rio de Janeiro*
[1]*Cuba*
[2]*Brazil*

1. Introduction

"Air Pollution" means the introduction by man, directly or indirectly, of substances or energy into the air resulting in deleterious effects of such a nature as to endanger human health, harm living resources and ecosystems and material property and impair or interfere with amenities and other legitimate uses of the environment, and "air pollutants" shall be construed accordingly (18 ILM 1442, 1979[1]).

Among the main reasons of increased air pollution is the expanding use of fossil energy sources, particularly in energy facilities, like the present case. Under the distributed power generation program that is currently being implemented in the country, generator sets (GS) have been installed and are still being installed in urban and sub-urban areas, to generate electricity using fossil fuels, such as in base-load as in emergency cases. During the operation of this equipment pollutants are released to the atmosphere, primarily nitrogen oxides (NO_x), sulphur dioxide (SO_2) and particulate matter (PM), which increase the concentrations of these pollutants in the atmosphere, affecting air quality.

It is essential to control these emissions so that concentrations achieved by each of the released pollutants be below their regulated values; therefore it is necessary to know to what extent air concentrations resulting from emissions from each facility could change. To this end, direct measurements and estimates of their concentrations by using Air Quality Models could be made. The measurements require specialized instrumentation and trained personnel, which increases the cost of the task; however, the use of models is much cheaper. They present algorithms to simulate physical and chemical processes for each pollutant in the atmosphere, allowing us to evaluate their behaviour after them being released. Despite this fact, none of the two methods is self-sufficient; so, it is necessary them be complemented. In many countries, particularly in developing countries, there are insufficient resources to carry out measurements of air quality with the necessary temporal and spatial extent, thus the computational modelling using the Air Quality Models carries the most importance. Moreover, the use of models for assessing air quality, and possible mitigation measures considering the results, often fall on simplified models due to their few requirements of input data.

The implementation of sophisticated models enables a more accurate assessment, however, their requirements and data complexity grow, and they are not always available in such countries. To this end, solutions to overcome the disadvantage of lack of data were taken.

The Environmental Protection Agency (US EPA) established on October 21st, 2005 the AERMOD modelling system as the model recommended to be used for the dispersion of local pollutants, replacing ISCST3 (Industrial Source Complex Short Term version 3), hitherto used.

AERMOD is a steady state, Gaussian local model, which includes the treatment of surface and elevated sources, both in simple and complex field (Fonseca, 2010)[2]. It is fed with surface hourly meteorological data and upper air. It is used in many countries in accordance with the regulations. The solutions taken in Cuba allow the use of the model even when the upper air data are not available, based only on surface data (Turtós et al., 2010)[3].

CALMET-CALPUFF Modelling System was developed by Earth Tech (Concord, MA) and it is the model proposed since 2003 by the U.S. EPA as the regulatory model to be used to perform detailed modelling of air pollution dispersion processes, in regional domains (at distances between 50 and 200 km from the source, with acceptable values up to 300 km) using three-dimensional wind fields. It is also proposed to be used at local level (from 0 to 50 km away from the source) in case of complex meteorological conditions such as those arising from the presence of hills and large bodies of water within the study area.

CALPUFF is a multi-layer, multi-species, non-steady state puff dispersion model which can simulate the effects of time- and space-varying meteorological conditions on pollutant transport, transformation, and removal. CALPUFF can use the three-dimensional meteorological fields modelled by the CALMET model, or simple, single station winds in a format consistent with the meteorological files used to drive the ISCST3, AUSPLUME or the CTDMPLUS steady-state Gaussian models. However single-station ISCST3, CTDMPLUS or AUSPLUME winds do not allow CALPUFF to take advantages of its capabilities to treat spatially-variable meteorological fields. CALPUFF contains algorithms for near-source effects such as building downwash, transitional plume rise, partial plume penetration, and subgrid scale terrain interactions as well as longer range effects such as pollutant removal (wet scavenging or dry deposition), chemical transformation, visibility effects (Scire et al., 2000)[4]. As the AERMOD, CALMET also requires upper air meteorological data, although the solutions taken in Cuba for AERMOD have not been introduced in CALMET given the complexity in introducing the meteorological data grid. Therefore, it is proposed to feed CALMET with results from the pre-processors of mesoscale meteorological and geophysical data, such as WRF and MM5.

The present chapter assesses the ability of the CALPUFF modelling system to simulate the dispersion of pollutants at local level, instead of using AERMOD, and predict the maximum concentrations each pollutant could achieve. CALPUFF was selected for the local domain as it provides more accurate results under complex meteorological conditions resulting from Cuban modelling domain, since it uses a three-dimensional meteorological grid, thus obtaining a better approximation for each variable as the resolution of each cell is higher. However, the AERMOD (EPA-454/B-03-001)[5] - the most advanced of those used in Cuba and regulatory EPA for local domains – assumes meteorological conditions to be uniform throughout the domain, what is a potential source of error in the model's input data. Other advantages of the CALPUFF is that it takes into account, although not at the same level of complexity as photochemical models do, the reactions occurring among pollutants in the atmosphere, making it possible to obtain the values of concentrations and deposition of sulphate and nitrate aerosols.

2. Background

As a result of the decision to install GS fed with fuel oil in Havana, a study of the related impacts to air quality that would produce the operation of the 11 planned GS by using the ISCST3 model (Turtós et al, 2006)[6] was carried out. This study concluded that significant amounts of oxides of sulphur and nitrogen will be released into the atmosphere with the simultaneous operation of these devices in a densely populated area, according to data provided by the manufacturer. In the selected modelling domain of 50 x 37 km, average population density is approximately 1240 inhabitants per km2, according to the data of the year 2000 from Population Study Center.

The work showed concentration levels of SO_2 and NO_2 to be achieved in 1850 points of Havana as a result of simultaneous operation of all the generator sets, which exceeds in an appreciable number of receptors the regulated values in the country for one-hour and one-day periods, and even permissible values from the World Health Organization (WHO).

As a result, it was decided to install only five GS with higher stack (from 15 to 37.5 m) and to carry out several studies (INEL, 2008[7]; CUBAENERGIA, 2009[8] and CUBAENERGIA, 2010[9]). The 2009 study was divided into two parts; the first was a comparative study between ISCST3 and AERMOD, and the second one, the modelling of different scenarios by using AERMOD. The 2010 study included measurement of emissions in different sites and modelling with AERMOD. Although five sites were built, only four are finally in operation.

The use of the AERMOD has been questioned since it is a domain that includes a coastline, which is not properly reflected by the uniform meteorology used by the model. Despite its high computational cost, the use of CALPUFF in this case is justified.

3. Methodology

CALPUFF calculates the pollutant dispersion in the receptor sites, taking into account complex three-dimensional wind fields, obtained with CALMET, which is particularly important for emission sources located in coastal areas and near high elevations. CALMET calculates wind structures in the study area from surface and upper air data. WRF outputs - model for calculating and forecasting the meteorological variables- was used as an alternative due to the unavailability of upper-air meteorological data in Cuba.

This system calculates the concentrations in the receptors distributed in the domain of study, at different times, for example, PM_{10}, sulphur dioxide (SO_2) and the species of the nitrogen family (NO_x). It also includes a simple model of chemical transformation to study and calculate some minor species such as sulphates (SO_4^{-2}) and nitrates (NO_3^-), which have a lot of relevance because of their potential effects on human health. Therefore, the system has been used in studies as a conceptual basis for these tests (Fonseca, 2010).

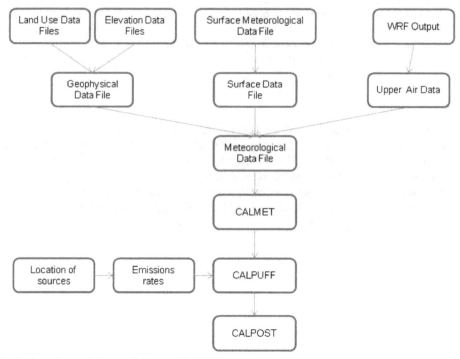

Fig. 1. Flowchart of the modelling with CALPUFF.

3.1 Reference values

The maximum permissible concentrations according to the Cuban Standard (NC) (NC 39:1999[10] and NC 111-2002[11]) and the reference values of the World Health Organization (WHO, 2005)[12] are used as reference for the analysis of results. These concentration values should not be higher than those of the pollutant (in the atmosphere) in the established time interval.

Maximum permissible concentrations ($\mu g/m^3$)		
Pollutant	1 hour	24 hours
SO_2	500	50
NO_x	85	40
PM_{10}	100	50

Table 1. Maximum permissible concentrations ($\mu g/m^3$) according to the Cuban Standard (NC) and the World Health Organization (WHO).

4. Case study: Modelling scenarios

In this chapter, we used the Air Quality Model CALPUFF in the evaluation of measures related to the design and location of the GS in order to mitigate the possible effects caused by these technologies on air quality.

Two scenarios were defined, which underwent a thorough analysis. The influence of meteorological conditions was assessed in detail, as well as the effective height of emission and power of each facility from a scenario to another.

Scenarios:

* Scenario 1 or proposed scenario (an initial project with 11 GS to be installed).
* Scenario 2 or real scenario (with 4 GS actually installed).

4.1 Modelling period

The modelling period (approximately two months of calculation) was that between January 2nd, 2009 at 00:00 pm and March 7th, 2009 at 23:00 pm.

4.2 Geophysical and meteorological data required for meteorological processing by using CALMET

The characteristics of geophysical and meteorological data used by CALMET for meteorological processing are set out below.

4.2.1 Topography and land use

In order to get information on the relief Digital Elevation Models (DEM) were used while land uses were obtained from the Global Database of Soil Coverage features, both available on the Internet.

4.2.2 Surface meteorological data

Hourly meteorological data of the entire modelling period were processed. They were collected in the seven surface stations located within the meteorological domain. See Table 2.

Station Code	Name	Coordinates	
		Latitude	Longitude
78322	Batabano	22.717° N	82.267° W
78318	Bahia Honda	22.92° N	83.17° W
78325	Casablanca	23.167° N	82.350° W
78373	Santiago de las Vegas	22.967° N	82.367° W
78320	Güira de Melena	22.78° N	82.52° W
78375	Melena del Sur	22.767° N	82.117° W
78376	Bauta	22.967° N	82.517° W

Table 2. Location of surface stations.
(*Source: INSMET, 2008*)

4.2.3 Upper air data – WRF configuration

WRF outputs are used in order to feed CALMET with upper air data. The WRF model was implemented with the configuration shown in the following tables.

Main Data	Domain 1	Domain 2	Domain 3
Cells	45 * 30	34 * 34	34 * 34
Cell Size	27km	9km	3km
Center location (Lambert Conformal Conic)	22.19°N 79.52°W	23.1°N 82.35°W	23.1°N 82.35°W

Table 3. Modelling domain.

Physical Data	Domain 1	Domain 2	Domain 3
Microphysics	WSM5 (A more sophisticated version of WSM3, allows for mixed-phase processes and super-cooled water) (mp_physics=4)	WSM3	WSM3
Cumulus Parameterization	Kain-Fritsch scheme (Deep and shallow convection sub-grid scheme) (cu_physics=1)	Kain-Fritsch scheme (Deep and shallow convection sub-grid scheme) (cu_physics=1)	(Not necessary for domains with cells lower than 4 km)
Shortwave Radiation	RRTMG (a shortwave scheme with Montecarlo Integrated Column Approach (MCICA) method of random cloud overlap) (ra_sw_physics=4)	Dudhia Scheme	Dudhia Scheme
Longwave Radiation	RRTMG (new RRTM scheme that includes MCICA method of random cloud overlap) (ra_sw_physics=4)	RRTM (Rapid Radiative Transfer Model)	RRTM (Rapid Radiative Transfer Model)
Surface Layer	MM5 similarity	MM5 similarity	MM5 similarity
Land Surface	5-layer thermal diffusion	5-layer thermal diffusion	5-layer thermal diffusion
Planetary Boundary layer	Yonsei University scheme	Yonsei University scheme	Yonsei University scheme

Table 4. Physical parameters.

The domains 1, 2 and 3, mentioned in the previous tables, in which the WRF model was implemented, are shown below.

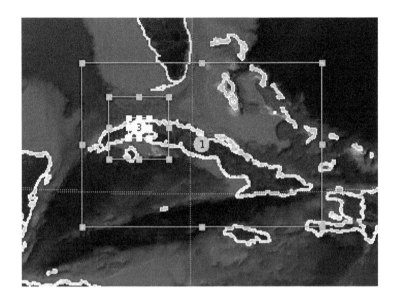

Fig. 2. Domains for WRF modelling.
(*Source: Capote et al. n.d.*)[13]

4.3 Structuring the CALMET meteorological grid

A rectangular grid of 90 x 95 km with a 1 km resolution and centered in X = 340500 Y = 340800 m was set up.

The following values collected by the surface stations within the meteorological domain were used: values of speed and prevailing wind direction, temperature, cloud coverage, height of the cloud base, pressure, relative humidity and precipitation rate. Upper air data obtained from processing WRF and adapted to the needs of CALMET by CALWRF (CALWRF, 2008)[14] pre-processor were added to these data. The description of these meteorological variables used by CALMET was obtained for 10 intervals of different heights (0 to 20, 20 to 40, 40 to 80, 80 to 160, 160 to 320, 320 to 640, 640-1200, 1200 to 2000, 2000 to 3000 and 3000 to 4000 m above ground level), thus obtaining the three-dimensional meteorological grid.

The CALMET meteorological field was simulated for a 1-hour time scale and was obtained by intelligent interpolation mechanisms applied to all the above variables.

Fig. 3 shows a sub-domain of the wind field obtained from CALMET. Examples of the wind direction and speed variability in an area of the modelled domain are given.

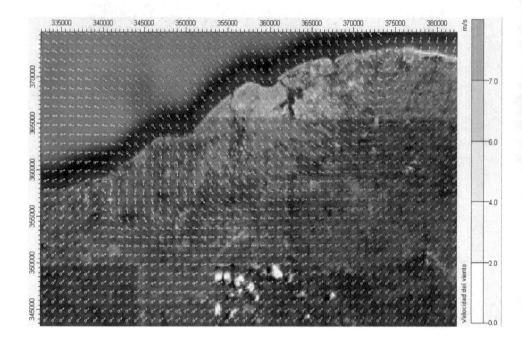

Fig. 3. Speed and direction of winds to 240 m above ground level.
(*Source: Base Map Google Earth*)

The arrows illustrate the wind direction (coinciding with that of the arrows) and speed
(color scale) in a study sub-domain.

4.4 Processing by CALPUFF

For processing with CALPUFF two more grids were defined - the calculation and receptor
grids - which are made to coincide with the meteorological grid, so that the puff modelling
cover a larger area in order to obtain the values of the concentrations and deposition or
removal flow of released pollutants. In the calculation grid, puffs are released and
transported, allowing their dispersion modelling, and it is in the receptors where the final
values of each study are obtained. A total of 8550 receptors, arranged in the shape of a
rectangular grid at a distance of 1 km from receptor to receptor, resulted from this
configuration.

Air Quality Study, Comparison Between the Proposed and Actual Scenarios of Generator Sets in Havana, by Using CALPUFF Model

105

Calculations were made using a pseudo-first order chemical reaction mechanism for the conversion of SO_2 to SO_4^{-2} and NO_x to nitrate aerosols. This mechanism is based on the chemical transformation scheme used in the MESOPUFF II model (Scire et al., 1984)[15], which introduces the most significant dependencies of transformation rates over varying environmental conditions, in time and space. This scheme models 5 species (SO_2, SO_4^{-2}, NO_x, HNO_3 y NO_3^{-1}) to which the modelling of suspended particles with diameter less than 10 micrometers (PM_{10}) was added.

5. Proposed scenario

Behold the initial project for the installation of 11 in Havana. See Fig. 4.

Fig. 4. Proposed location of the 11 generator sets to be initially installed in Havana. (*Source: Base Map Google Earth*)

The following tables show the composition of each of the GS to be installed, their technical features and emission rates.

Emission values for the study were those provided by the manufacturer. A fuel with a sulphur content of 2% was assumed to be used.

Nu.	Name of GS	Number of engines making them up	Composition of the GS	Number of stack	Engine power unit (MW)
1	Guanabacoa (GUA)	16	4 x 4	4	1.7
2	Apolo (APOLO)	16	4 x 4	4	1.7
3	Naranjito (NAR)	16	4 x 4	4	1.7
4	Victoria de Giron (GIRON)	16	4 x 4	4	1.7
5	Diezmero (DIEZ)	16	4 x 4	4	1.7
6	San Agustin (SANAG)	16	4 x 4	4	1.7
7	Regla (REGLA)	28	7 x 4	7	1.7
8	Cotorro (COTO)	24	6 x 4	6	2.5
9	Parque Metropolitano (PMA)	24	6 x 4	6	2.5
10	Berroa (BERROA)	24	6 x 4	6	2.5
11	CUJAE (CUJAE)	24	6 x 4	6	2.5
	Total	220		55	450.8

Table 5. Composition of GS to be installed.

Nu.	Name of GS	Stack Height (m)	Stack Diameter (m)	Output Speed (m/s)	Output Temperature (°K)
1	NAR	15	1.27	14.98-11.23*	520
2	SANAG	15	1.27	14.98-11.23*	520
3	APOLO	15	1.27	14.98-11.23*	520
4	CUJAE	15	1.27	23.1-17.32*	553.5
5	REGLA	15	1.27	14.98-11.23*	520
6	BERROA	15	1.27	27.7	504.15
7	COTO	15	1.27	27.7	504.15
8	DIEZ	15	1.27	18.1	477.15
9	GIRON	15	1.27	18.1	477.7
10	GUA	15	1.27	18.1	477.7
11	PMA	15	1.27	27.7	504.15

* In these cases, it was considered that the output speeds change with variations in emission rates.

Table 6. Specifications of the stacks of the GS to be installed.

LCC North Cuba geographic projection (LCC-CN) was used, since it meets the study's needs. The LCC-CN projection parameters are:

- Projection Origin (22.35° of north latitude and 81° of west longitude)
- Standard Parallels (21.7 ° and 23 ° North)
- False North and False East (X = 500000 Y = 280296.016 m)

Nu.	Name of GS	$SO_2(g/s)$	$NO_x(g/s)$	$PM_{10}(g/s)$
1	NAR	12.8	13.84	0.1932
2	SANAG	12.8	13.84	0.1932
3	APOLO	12.8	13.84	0.1932
4	CUJAE	18.84	19.44	0.6906
5	REGLA	12.8	13.84	0.1932
6	BERROA	18.84	19.44	0.6906
7	COTO	18.84	19.44	0.6906
8	DIEZ	12.8	13.84	0.1932
9	GIRON	12.8	13.84	0.1932
10	GUA	12.8	13.84	0.1932
11	PMA	18.84	19.44	0.6906

Table 7. Emission values for each source (g/s).

Maximum emissions for each engine (by its power) are showed, though it is considered that each set worked with 87.5% availability.

5.1 Analysis of results and comparison with those of the reference values

Table 8 shows the maximum modelled concentrations by each pollutant species in the atmosphere for each of the intervals of importance in the study.

	Maximum concentrations		
Pollutant	1 hour	24 hours	the whole period (1559 hours)
SO_2	558	168	53
SO_4^{-2}	8.1	2.4	0.1
NO_x	567	180	56
NO_3^{-1}	27	3.5	0.2
PM_{10}	21	5.4	0.9
HNO_3	23	3.8	0.6

Table 8. Maximum modelled concentrations ($\mu g/m^3$) for each pollutant for each time interval.

It is possible to check if the standardized values are exceeded by comparing the maximum modelled concentrations by each species in 1-hour and 24-hour periods with the maximum permissible concentrations (for 1 and 24 hours respectively) according to Cuban standards. It should be noted that there are species for which the standards do not provide maximum permissible concentrations.

Table 9 below shows a ratio between reached maximum concentrations and maximum permissible (for each species and time interval). When this value is greater than 100, then this species exceeds its maximum permissible for that interval of time.

Pollutant	MC / MPC (%) 1 hour	24 hours
SO_2	112	336
NO_x	667	450
PM_{10}	21	11

Table 9. Ratio of the maximum Modelled Concentrations (MC) and the Maximum Permissible Concentrations (MPC) according to the NC and WHO, for different time intervals.

Taking into account the above criteria it can be confirmed that both SO_2 and NO_x (values highlighted in red) go over the MPC in both evaluation periods.

The following table shows the maximum average concentrations reached by the two "critical species".

Pollutant	Maximum average concentrations 1 hour	24 hours
SO_2	66.2	11.9
NO_x	68.8	11.6

Table 10. Maximum average concentrations of SO_2 and NO_x ($\mu g/m^3$).

5.1.1 Critical receptors

The following tables show the receptors, in which the maximum concentrations of SO_2 and NO_x are obtained and at the same time exceed the MPC:

Evaluation Period	NuO	Number of Receptors
1 hour	1	3
	1	31
	2	13
	3	9
	4	3
24 hours	5	3
	8	1
	9	1
	10	1
	15	1
	32	1

Table 11. Number of receptors in which SO_2 exceeds the MPC and number of opportunities (NuO) this occurs.

Table 11 shows the number of opportunities SO_2 exceeds its MPC (middle column), as well as the amount of receptors in which this occurs (right column).

Analyzing the maximum hourly concentrations of NO_x, it is observed that exceeds its MPC for 1-hour and for 24-hour periods in many receptors, therefore only the most critical receptors are shown in Table 12. It is worth noting that 687 receptors go over the MPC for 1-hour period, while the most critical receptor exceeds this MPC for 315 hours out of the 1559 modelled hours, which represents approximately 20% of the modelling period.

Evaluation Period	NuO	Number of Receptors
1 hour	112	2
	132	2
	138	1
	149	1
	155	1
	185	1
	188	1
	315	1

Table 12. Number of receptors that NOx exceeds the MPC for 1-hour period (middle column) and the NuO it occurs (right column).

Evaluation Period	NuO	Number of Receptors
24 hours	1	68
	2	20
	3	13
	4	5
	5	6
	6	6
	7	2
	8	2
	10	3
	12	3
	13	1
	18	1

Table 13. Number of receptors that NOx exceeds the MPC for 24-hour period (middle column) and the NuO it occurs (right column).

The following figures show the isolines of the maximum daily and hourly concentration of NO_x and SO_2 respectively.

The sources are identified by red crosses and all locations where the NOx exceeds its MPC by red areas.

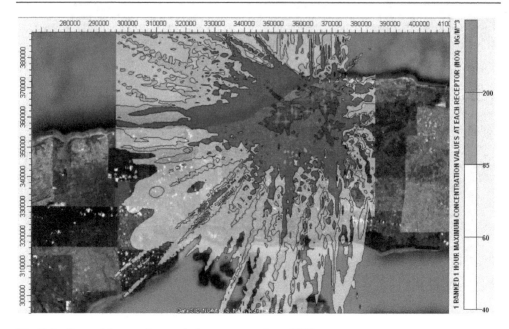

Fig. 5. Isolines of the maximum daily concentration of NO$_x$.
Source: Base Map Google Earth)

Fig. 6. Isolines of the maximum hourly concentration of SO$_2$.
(*Source: Base Map Google Earth*)

5.1.2 Depositions

There is general consensus that the deposition of sulphur compounds and nitrogen causes acidification on aquatic and terrestrial ecosystems, which means, among other impacts, less fertile soils and impacts to aquatic organisms which can not tolerate the acidity conditions. In general, these effects appear when the threshold of the critical load is exceeded. The critical load has been defined as "the maximum deposition of acidifying compounds that will not cause chemical changes leading to long term harmful effects on ecosystem structure and function" (Sverdrup et al., 1995)[16].

As there are not critical load values for Cuban conditions, a comparison could be made with the values set in the Air Quality Guide for Europe, WHO, 2000, shown in Table 14.

Compound	Guide values of critical loads	Evaluation Period
S	250–1500 eq/ha/year*	annual
N	5–35 kg N/ha/year*	annual

* Depending on soil type and ecosystem

Table 14. Values set in the Air Quality Guide for Europe.

Table 15 shows the maximum and average deposition values of sulphur and nitrogen, which are considered valid in practice, expressed in $\mu g/m^2$ for each evaluated period, and their respective conversions to eq / ha and to kg N / ha, depending on the species that identify the critical loads for each compound.

Evaluation Period		N		S	
		(µg/m2)	(Kg_N/ha)	(µg/m2)	(eq/ha)
1 hour	Maximum	2490	---	5190	---
	Average	77	6.7	355	1944
24 hours	Maximum	11600	---	23000	---
	Average	327	1.2	1520	347
1559 hours	Maximum	111000	6.2	203000	713
	Average	2830	0.2	11900	42

Table 15. Maximum and average deposition values of sulphur and nitrogen.

The table above shows that nitrogen does not exceed critical load values, while the average sulphur for 1-hour period does exceed them.

6. Real scenario

It is the one that takes into account the four GS currently installed in Havana.

Fig. 7. Location of the four generator sets currently installed in Havana.
(*Source: Base Map Google Earth*)

The following table shows the composition of each of the currently installed GS.

Nu.	Name of GS	Number of engines making them up	Composition of the GS	Number of Stack	Engine power unit (MW)
1	APOLO	16	4 x 4	4	1.7
2	SANAG	16	4 x 4	4	1.7
3	REGLA	28	7 x 4	7	1.7
4	CUJAE	24	6 x 4	6	2.5
	Total	84		21	162

Table 16. Composition of the 4 GS currently installed in Havana.

The technical data of each of the stacks of these 4 GS match those assumed in the initial project, except for the height, since they are 37.5 m. See Table 6.

The emission values of each of these sources are the same as those used in the proposed scenario, since the values provided by the manufacturer were used in both studies. See Table 7.

Air Quality Study, Comparison Between the Proposed and Actual Scenarios of Generator Sets in Havana, by Using
CALPUFF Model

113

6.1 Analysis of results and comparison with the reference values

Table 17 shows the maximum concentrations modelled in the atmosphere for each pollutant species for each of the intervals of importance in the study.

Maximum concentrations			
Pollutant	1 hour	24 hours	All Period (1559 hours)
SO_2	435	105	38
SO_4^{-2}	4.5	1.11	0.045
NO_x	441	112	40
NO_3^{-1}	12.6	1.9	0.068
PM_{10}	16	3.4	0.57
HNO_3	18.7	2.14	0.3

Table 17. Maximum concentrations reached ($\mu g/m^3$) by each pollutant for each time interval.

It is possible to check if the standardized values are exceeded by comparing MC with their respective MPC according to Cuban standards.

Table 18 below shows a ratio between MC and the MPC (for each species and time interval), following the same criteria as in the previous scenario, where this value is greater than 100, then this species exceeds its MPC for that time interval.

MC / MPC (%)		
Pollutant	1 hour	24 hours
SO_2	87	210
NO_x	519	280
PM_{10}	16	6.8

Table 18. Ratio between MC and MPC according to NC and WHO, for different time intervals.

Given the above criteria, it can be confirmed that both SO_2 for 24-hour periods and NO_x for 1-hour and for 24-hour periods exceed their MPC (values in red).

The following table shows maximum average concentrations modelled in the entire domain compared to the previous scenario.

Maximum average concentrations		
Pollutant	1 hour	24 hours
SO_2	33.9	4.5
NO_x	35.5	4.39

Table 19. Maximum average concentrations of SO_2 y NO_x ($\mu g/m^3$).

6.1.1 Critical receptors

The following tables show the receptors in which NO_x and SO_2 exceed the MPC and the number of opportunities it happens.

Table 20 shows the number of opportunities the SO_2 exceeds its MPC daily (middle column) and the amount of receptors in which this happens (right column).

Evaluation Period	NuO	Number of Receptors
	1	6
	2	2
24 hours	3	2
	6	1
	25	1

Table 20. Number of receptors in which SO_2 exceeds the MPC and the number of opportunities (NuO) it occurs.

After analyzing the maximum hourly concentrations of NO_x, it can be observed that it exceeds its MPC in many receptors, therefore only the most critical ones will be showed, although it is worth noting that 207 receptors go above MPC for 1-hour period, while the most critical receptor exceeds this MPC for 251 hours out of the total modelled, representing approximately 16% of the modelling period.

Evaluation Period	NuO	Number of Receptors
	42	1
	49	1
	59	1
	64	1
1h	77	1
	90	1
	92	1
	105	1
	251	1

Table 21. The most critical receptors where NOx exceeds for 1-hour period maximum permissible concentrations and the number of opportunities this happens.

The following table shows the days in which MPC are exceeded by NO_x

Evaluation Period	NuO	Number of Receptors
	1	13
	2	1
	3	7
24 hours	5	1
	6	1
	7	1
	32	1

Table 22. The most critical receptors where NOx exceeds in 24 hours maximum permissible concentrations and the number of opportunities this happens.

The following figures show the isolines of the maximum daily and hourly concentration of NO_x and SO_2 respectively.

The sources are identified by red crosses and all locations where the NOx exceeds its MPC by red areas.

Air Quality Study, Comparison Between the Proposed and Actual Scenarios of Generator Sets in Havana, by Using
CALPUFF Model

115

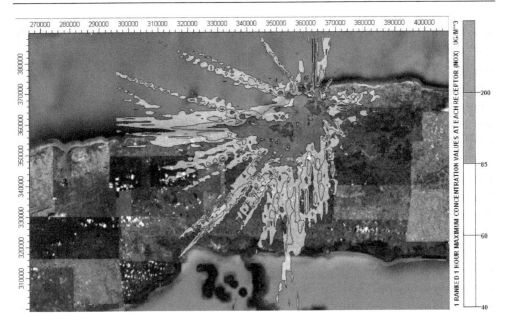

Fig. 8. Isolines of the maximum daily concentration of NO$_x$.
(*Source: Base Map Google Earth*)

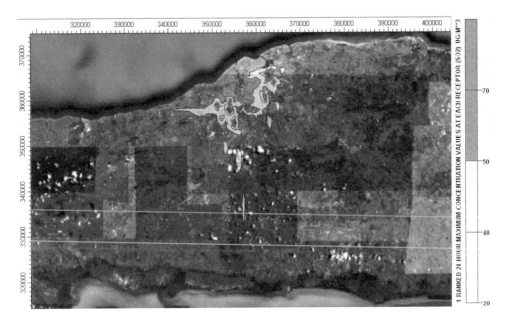

Fig. 9. Isolines of the maximum hourly concentration of SO$_2$.
(*Source: Base Map Google Earth*)

6.1.2 Deposition

Table 23 shows the maximum and average deposition values of sulphur and nitrogen, which are considered valid in practice, expressed in $\mu g/m^2$ for each evaluated period, and their respective conversions to eq / ha and to kg N / ha, depending on the species that identify the critical loads for each compound.

Evaluation Period		N ($\mu g/m^2$)	(Kg_N/ha)	S ($\mu g/m^2$)	(eq/ha)
1 hour	Maximum	1500	----	3180	----
	Average	40	3.5	185	1013
24 hours	Maximum	8680	---	17500	---
	Average	131	0.5	568	130
1559 hours (All period)	Maximum	83200	4.7	154000	541
	Average	884	0.05	3500	12.3

Table 23. Maximum and average deposition values of sulphur and nitrogen.

The above table shows that neither nitrogen nor sulphur exceeds their critical load values.

7. Comparison between scenarios

The following table shows a ratio between the maximum modelled concentrations by each species in the Scenario 1 (MC1) and the Scenario 2 (MC2).

Pollutant	1 hour	24 hours	MC1/MC2 the whole period (1559 hours)
SO_2	1.28	1.6	1.39
SO_4^{-2}	1.8	2.16	2.22
NO_x	1.29	1.61	1.4
NO_3^{-1}	2.14	1.84	2.94
PM_{10}	1.31	1.59	1.58
HNO_3	1.23	1.78	2

Table 24. Ratio between MC1 and MC2.

The above table shows that all the MC decrease between 1.23 and 2.94 times.

If the ratio between MC1 and MC2 with respect to MPC (see Table 9 y Table 18) is analyzed, it can be observed that SO_2 and NOx exceed their MPC for 1-hour and for 24-hour periods in the scenario 1, while in the scenario 2, NOx exceeds their MPC for 1-hour and for 24-hour periods, but SO_2 only exceeds its MPC for a 24-hour period.

As the MC only show the behaviour in the most critical receptor and there is not a complete analysis of the entire domain, the ratio between MC1 and MC2 really does not show the variation from one scenario to another. Therefore, maximum average concentrations were calculated to qualitatively get a better idea of the critical pollutant dispersion in every scenario (see Table 10 y Table 19). As a result there was a decrease of about half in the concentrations.

The decrease in absolute maximum hourly concentrations of the 2 critical species is mainly due to the elimination of the GS responsible for them, since they are caused by the action of individual sites and not by overlapping plumes of several sites, for example: the maximum hourly concentration of SO_2 and NO_x in the scenario 1 is reached in the vicinity of the COTO site, while it is achieved near CUJAE in the scenario 2. However, the overall decrease in maximum average concentrations is due to the increase of 22.5m in the height of the stacks in the four currently installed GS. For example, the maximum daily concentration of SO2 reached in the vicinity of the REGLA site in the scenario 1 is 168 µg/m³, while in the scenario 2 decreases to 105 µg/m³.

Analyses carried out to deposition levels showed that only the average sulphur for 1 hour in the scenario 1 is the one which exceeds the critical load values for Europe. See Table 15 y Table 23.

As the critical load depends on the past and present management, of the type of ecosystem and soil conditions, to what extent these critical load values proposed for Europe can be applied in other regions is not known. The information required to properly analyze these results and turn them into physical impacts and external costs is difficult to obtain.

In order to qualitatively compare how much the level of deposition from one scenario to another decrease, ratio of the maximum and average deposition of sulphur and nitrogen in each evaluation period was analyzed.

1 hour	Maximum	1.7	1.6
	Average	1.9	1.9
24 hours	Maximum	1.3	1.3
	Average	2.5	2.7
1559 hours	Maximum	1.3	1.3
	Average	3.2	3.4

Table 25. Ratio of the maximum and average deposition of sulphur and nitrogen in each evaluation period from one scenario to another.

The table above shows that maximum deposition levels decreases between 1.3 and 1.7 times, while average deposition levels decreased between 1.9 and 3.4 times.

As with maximum concentrations, the highest hourly deposition in the scenario 1 is observed near COTO site and in the scenario 2 - in the vicinity of CUJAE site. This is also due to the elimination of the site where the maximum emissions occur.

8. Comparison between AERMOD and CALPUFF

The real scenario was modelled using the AERMOD (the same period, the same sources, emission rates, etc.). Results, very similar to those obtained with CALPUFF for the same scenario, were obtained. Because of the similarities in the results, we will analyze only the behaviour of SO_2 in the REGLA site.

Table 26 shows the behaviour of some of the maximum concentrations in the vicinity of the REGLA.

Period	Model	Concentration of SO_2 ($\mu g/m^3$)
1 hour	CALPUFF	339
	AERMOD	312
24 hours	CALPUFF	93
	AERMOD	141

Table 26. Maximum concentrations of SO_2 ($\mu g/m^3$) in the vicinity of **REGLA**.

The above table shows that the results are comparable.

This is only a preliminary analysis due to the short temporal modelled interval, but it shows the possibility of using the AERMOD in the evaluation of local pollutant dispersion in Havana despite the existence of the extensive coastline. In order to obtain a final result, an assessment of air quality for a longer modelling period (at least for a year) is recommended to be carried out since this makes it more likely to show all possible synoptic conditions, which does not occur in two-month period.

9. Conclusions

- A refined modelling was performed using the CALPUFF model, following the recommendations for using it at the local level under complex weather conditions. This is considered a significant progress because the AERMOD was the model used for national studies so far.
- The WRF-CALMET-CALPUFF methodology was first used in a real case.
- Significant reductions in air pollution were obtained by replacing the initial project (proposed scenario) with the end project (real scenario) for the installation of generator sets, regarding the following aspects:
  1. Decrease in maximum concentrations with the elimination of the site that produced them, i.e. decrease in absolute maximum hourly concentrations of SO2 and NOx by removing the **COTO**.
  2. Decrease in maximum concentrations by increasing height of the stacks, i.e. decrease in absolute maximum daily concentrations of SO_2 of **COTO**.
  3. Decrease in maximum concentrations by increasing height of the stacks, i.e. decrease in absolute maximum daily concentrations of SO_2 of GS **COTO**.
- Despite the improvements in air quality when using the scenario 2, actions must be taken and new alternatives should be developed to continue reducing emissions so that the maximum concentrations of these pollutants do not exceed their maximum permissible concentrations.
- It was found that the results obtained with CALPUFF and AERMOD, despite the differences between the models and short modelling period, provide comparable results in assessing the dispersion of pollutants at local scale for scenarios in Havana, what makes it possible the use of the AERMOD instead of the CALPUFF model in such scenarios.

Air Quality Study, Comparison Between the Proposed and Actual Scenarios of Generator Sets in Havana, by Using
CALPUFF Model

119

- It is recommended to carry out a future local study in Havana (at least one-year assessment) in order to obtain conclusive results about the similarities by using one or the other model in these scenarios.

10. References

1 18 ILM 1442 (1979). Convention on Long-Range Transboundary Air Pollution, Article 1(a).

2 Fonseca,Y. (2010). *Implementación y aplicación del sistema de modelación CALMET-CALPUFF-CALPOST a escala local.* Trabajo de Diploma. Instituto Superior de Tecnologías y Ciencias Aplicadas (INSTEC), Departamento de Ingeniería Nuclear.

3 Turtós L., J. Rivero, L. Curbelo, M. S. Gácita, E. Meneses & N. Díaz, 2010. *Methodological guide for implementation of the AERMOD system with incomplete local data. Atmospheric Pollution Research,* ISSN: 1309-1042, Vol 1, Issue 2, pp 102-111.

4 Scire, J. (2000). *User's guide for the CALPUFF Dispersion Model,* July 2011, Available from: www.src.com/calpuff/download/CALPUFF_UsersGuide.pdf.

5 EPA-454/B-03-001, September 2004. *User's guide for the ams/epa regulatory model – AERMOD.* U.S. Environmental Protection Agency, Office of Air Quality Planning and Standards Emissions, Monitoring, and Analysis Division Research Triangle Park, North Carolina 27711

6 Turtós L, N. Díaz, M. Padrón & E. Molina, 2006, *Estudio de calidad del aire por la instalación de los Grupos Electrógenos en Ciudad de la Habana.* Informe de Servicio Científico Técnico entre CUBAENERGIA y la Empresa de Ingeniería para la Electricidad, INEL.

7 INEL, 2008. *Datos usados en el estudio global pronostico 2008.* Informe Técnico.

8 CUBAENERGIA, 2009. *Estudios ambientales vinculados a las emisiones de gases y material partículado total de los grupos electrógenos de fuel oil en Ciudad Habana.* Informe de Servicio Científico Técnico CUBAENERGIA-INEL.

9 CUBAENERGIA, 2010. *Informe de servicio de medición y modelación de dispersión de contaminantes atmosféricos emitidos por los grupos electrógenos de régimen base en La Habana.*

10 NC 39:1999, NORMA CUBANA. *Sistema de Normas para la protección del Medio Ambiente, Atmósfera, Requisitos higiénicos sanitarios: Concentraciones máximas admisibles, alturas mínimas de expulsión y zonas de protección sanitaria.* Comité Estatal de Normalización.

11 NC 111: 2002, NORMA CUBANA. *Calidad del aire, Reglas para la vigilancia de la calidad del aire en asentamientos humanos.* La Habana, Oficina Nacional de Normalización.

12 WHO, 2005, WHO/SDE/PHE/OEH/06.02. *Air quality guidelines for particulate matter, ozone, nitrogen dioxide and sulfur dioxide Global update 2005.* Summary of risk assessment.

13 Capote G., L. Turtós, L. Alvarez, A. Bezanilla & I. Borrajero, (n.d.).*Implementación del modelo meteorológico WRF en Cuba para su uso en modelos de dispersión de contaminantes atmosféricos.*

14 CALWRF, 2008. Version 1.1. Level 080429. *Atmospheric Study Group in TRC Environmental Corporation.*

[15] Scire et al., 1984. *User' Guide to the MESOPUFF II dispersion Model and related processor programs.* EPA-600/8-84-013. US Environmental Protection Agency. Research Triangle Park. NC.

[16] Air Quality Guidelines for Europe, 2000. World Health Organization. Regional Office for Europe, Copenhagen. WHO Regional Publications. European Series. No. 91. Second Edition.

6

Biomonitoring of Airborne Heavy Metal Contamination

Mehran Hoodaji, Mitra Ataabadi and Payam Najafi
Islamic Azad University, Khorasgan Branch (Isfahan)
Iran

1. Introduction

During the last few decades, heavy metal contamination of biotic component of environment has attracted the attention of many investigators. The main reason of these researches based on the heavy metal concentration may have a potential hazard in our food chain after a long period of procrastination. Using biological materials in the determination of environmental pollution as indicators is a cheap and reliable method. Various types of plant such as lichens, mosses, bark and leaves of higher plants, have been used to detect the deposition, accumulation and distribution of metal pollution (Akosy, 2008), Because of plants greatly affected by physical and chemical environmental conditions. If conditions become altered, the exposed plant community can accurately reflect these changes (Nash, 1988).

This chapter discusses the possibility of various types of plant usage as biomonitors for detection atmospheric heavy metal pollution in different conditions and factors that affect their accumulative potential.

1.1 Heavy metals

Over the past two decades, "heavy metals" has been used increasingly in various publications and in legislation related to chemical hazards and the safe use of chemicals. It is often used as a group name for metals and semimetals (metalloids) that have been associated with contamination and potential toxicity or ecotoxicity (Duffus, 2002).

Now, it is known that heavy metals represent a large group of chemical elements (> 40) with atomic mass > 50 carbon units (Fig 1.) Most of heavy metals may be important trace elements in the nutrition of plants, animals or humans (e.g. Zn, Cu, Mn, Cr, Ni, V), while others are not known to have positive nutritional effects (e.g. Pb, Cd, Hg). However all of these may cause toxic effects (some of them at a very low content level) if they occur excessively (Spiegel, 2002). The toxicity of heavy metal depends a great deal on their chemical form, concentration, residence time, etc (Mielke & Reagan, 1988). Because of these elements do not decay with time, their emission to the environment is a serious problem which is increasing worldwide due to the rapid growth of population, increasing combustion of fossil fuels, and the expansion of industrial activities (Smodis & Bleise, 2000).

1 H																	2 He
3 Li	4 Be				Metalloids							5 B	6 C	7 N	8 O	9 F	10 Ne
11 Na	12 Mg			Heavy Metals								13 Al	14 Si	15 P	16 S	17 Cl	18 Ar
19 K	20 Ca	21 Sc	22 Ti	23 V	24 Cr	25 Mn	26 Fe	27 Co	28 Ni	29 Cu	30 Zn	31 Ga	32 Ge	33 As	34 Se	35 Br	36 Kr
37 Rb	38 Sr	39 Y	40 Zr	41 Nb	42 Mo	43 Tc	44 Ru	45 Rh	46 Pd	47 Ag	48 Cd	49 In	50 Sn	51 Sb	52 Te	53 I	54 Xe
55 Cs	56 Ba	57 La	72 Hf	73 Ta	74 W	75 Re	76 Os	77 Ir	78 Pt	79 Au	80 Hg	81 Tl	82 Pb	83 Bi	84 Po	85 At	86 Rn
87 Fr	88 Ra	89 Ac	104 Rf	105 Db	106 Sg	107 Bh	108 Hs	109 Mt	110 Ds	111 Rg							

Lanthanides	58 Ce	59 Pr	60 Nd	61 Pm	62 Sm	63 Eu	64 Gd	65 Tb	66 Dy	67 Ho	68 Er	69 Tm	70 Yb	71 Lu
Actinides	90 Th	91 Pa	92 U	93 Np	94 Pu	95 Am	96 Cm	97 Bk	98 Cf	99 Es	100 Fm	101 Md	102 No	103 Lr

Fig. 1. Heavy metal position in Periodic Table.

1.1.1 Origin of heavy metals in the environment

There are two different sources for heavy metals in the environment. These sources can be both of natural or anthropogenic origin.

1.1.1.1 Natural

The principal natural source of heavy metals in the environment is from crustal material that is either weathered on (dissolved) and eroded from (particulate) the Earth's surface or injected into the Earth's atmosphere by volcanic activity. These two sources account for 80% of all the natural sources; forest fires and biogenic sources, account for 10% each. Particles released by erosion appear in the atmosphere as windblown dust. In addition, some particles are released by vegetation. The natural emissions of the some heavy metals are 12,000 (Pb); 45,000 (Zn); 1,400 (Cd); 43,000 (Cr); 28,000 (Cu); and 29,000 (Ni) metric tons per year, respectively (Nriagu, 1990).

1.1.1.2 Anthropogenic

There are a multitude of anthropogenic emissions in the environment. Generally heavy metals enter into the environment mainly via three routes: (I) deposition of atmospheric particulates (e.g. Mining, Smelting, Fossil fuel combustion, municipal waste incineration, cement production and phosphate mining). (II) disposal of metal enriched sewage sludge and sewage effluents, commercial Fertilizers and pesticides and animal waste specially to the terrestrial and aquatic environment (III) by-product from metal mining processes (Shrivastav, 2001; Smodis & Bleise, 2000). Among them, the major source of metals is from mining and smelting. Mining releases metals to the fluvial environment as tailings and to the atmosphere as metal-enriched dust whereas smelting releases metals to the atmosphere as a result of high-temperature refining processes (Adriano, 1986). The estimation of metal input into environment from the two latter sources (II & III) is relatively easy to measure, but atmospheric input is difficult to quantify accurately because of after emission, the pollutants are subjected to physical, chemical and photochemical transformations, which ultimately decide their fate depending upon the atmospheric concentrations (Shrivastav, 2001; Smodis & Bleise, 2000). Air pollutants do not remain confined near the source of emission, but spread over distances, transcending natural and political boundaries depending upon topography and meteorological conditions, especially wind direction, wind speed and vertical and horizontal thermal gradients (Smodis & Bleise, 2000). Among the various species present in the particulate matter, a great attention has been devoted since many years to the study of the elements with elevated toxicity and great diffusion in the environment (As, Cd, Cr, Hg, Ni, Pb, etc.) because of both anthropogenic and natural pollutant emissions (Beijer & Jernelöv, 1986). The contributions of natural and anthropogenic sources in atmospheric heavy metal pollution are shown in table 1.

Element	Fluxes from crustal origin	Fluxes from volcanic origin	Fluxes in gas of volcanic origin	PM emission of industrial origin	Element flux from fossil fuel	Total Emission	Atmospheric Alteration Factor
Ag	0.5	0.1	0.0006	40	16	50	8333
Al	356,500	132,750	8.4	40,000	32,000	72,000	15
As	25	3	0.1	620	160	780	2786
Cd	2.5	0.4	0.001	40	15	55	1897
Co	40	30	0.04	24	20	44	63
Cr	500	84	0.005	650	290	940	161
Cu	100	93	0.012	2200	430	2630	1363
Fe	190,000	87,750	3.7	75,000	32,000	107,000	39
Hg	0.3	0.1	0.001	50	60	110	27,500
Mn	4250	1800	2.1	3000	160	3160	52
Mo	10	1.4	0.02	100	410	510	4474
Ni	200	83	0.0009	600	380	980	348
Pb	30	8.7	0.012	16,000	4300	20,300	34583
Sb	9.5	0.3	0.013	200	180	380	3878
Se	3	1	0.13	50	90	140	3390
Sm	32	9	-	7	5	12	29
Sn	50	2.4	0.005	400	30	430	821
Ti	23000	12,000	-	3600	1600	5200	15
V	500	150	0.05	1000	1100	2100	323
Zn	250	108	0.14	7000	1400	8400	2346

Table 1. Natural and anthropogenic element fluxes (value $\times 10^8$ g y^{-1})
Atmospheric Alteration Factor (%) = Total Emission/ (Fluxes from crustal origin+Fluxes from volcanic origin)$\times 100$ (from Avino et al., 2008).

1.1.2 Fate and transport of atmospheric heavy metals in the environment

The environmental and human health effects of heavy metals depend on the mobility of each metal through environmental compartments and the pathways by which metals reach humans and the environment. The free ion is generally the most bioavailable form of a metal, and the free ion concentration if often the best indicator of toxicity. However, there are exceptions, such as the well known case of mercury, where the organic form, (methyl mercury) is more toxic than the inorganic ion. Metals exert toxic effects if they enter into biochemical reactions in the organism and typical responses are inhibition of growth, suppression of oxygen consumption and impairment of reproduction and tissue repair (Long et al., 1995). The World Health Organization (WHO) estimates that more than 2 million premature deaths occur annually worldwide, and these can be attributed to the effects of outdoor and indoor air pollution (WHO, 2002). The main outdoor air pollutants are as follows: particulate matter (PM), carbon monoxide, nitrogen oxides, sulfur oxides, ozone, and volatile organic compounds. Most of the mortality/morbidity caused by atmospheric pollution in urban areas is caused by PM with aerodynamic diameters less than 10 μm, known as the inhalable fraction (MacNee & Donaldson, 2000). Heavy metals occur, in atmosphere, basically in particulate form. PM-containing heavy metals may induce the oxidative stress mediated by reactive oxygen, with potential mutagenic effects as a strong correlation between the high elemental concentrations in aerosol particles and high mortality and morbidity has been found in several epidemiological studies (Manoli et al., 2002).

1.1.2.1 Mechanisms of heavy metal deposition

The aerosols, which have a very small falling velocity, are easily transferred by the wind and it is possible to be deposited through the rain at long distances from the point of their emission (Smirnioudi et al., 1998). Therefore, it is expected that chemical components in the rainwater (acid components, anions, cations and heavy metals) damage significantly the environment (surface waters, plants, animals, human beings). These metals even if deposited constantly in small rates over long periods of time, accumulate in the environment and will probably pose an increasing major environmental and human health hazard in future.

The transfer of airborne particles to land or water surfaces by dry, wet and occult deposition constitutes the first stage of accumulation of atmospheric heavy metals. The predominant path depends upon the type of chemical species and upon meteorological factors such as the intensity and distribution of rain fall.

1.1.2.1.1 Dry deposition

Dry deposition involves four distinct processes: gravitational settling, impaction, turbulent transfer and transfer by Brownian motion. The relative importance of these processes depends primarily on the size of particles (Shrivastav, 2001). Dry deposition is more likely to remove particulate forms. Two major categories of particulate matter are fine particles and coarse particles. Fine particulate matter (FPM) comprises particles with aerodynamic diameters of 2.5 μm or less. They are emitted from fossil fuel combustion, motor vehicle exhausts (including diesel) and wood burning. Several toxic metals, including arsenic, cadmium, lead, zinc, antimony and their compounds are associated with FPM in ambient air. This is important from a public health perspective since these fine particles are respirable and can be transported over very long distances. Coarse particulate matter or PM_{10} is the mass concentration of particulate matter having aerodynamic diameters less than 10 μm (USEPA, 1996). Anthropogenically added particulates and aerosols in atmosphere show a broad size distribution from 0.001 μm to 50 μm and are strongly influenced by atmospheric transport processes. In rural regions anthropogenic particles > 5 μm are absent (Ward et al., 1975) and in Polar Regions heavy metals are mostly associated with small aggregated particles in the range of 0.1-10 μm. Dry deposition of heavy metals in rural and remote regions is therefore, through impaction and turbulent transfer (Barrie &

Schemenauer, 1986). Dry deposition represents a major removal pathway for many pollutants from the atmosphere, and it is especially important in arid and semiarid regions where removal by wet deposition (i.e., rainfall scavenging) is greatly diminished because of limited precipitation (Seinfeld & Pandis 1998).

1.1.2.1.2 Wet deposition

In wet deposition, there are always some atmospheric hydrometeors which scavenge aerosol particles. This means that wet deposition is gravitational, Brownian and/or turbulent coagulation with water droplets. Wet deposition involves two major processes: nucleation and within and below cloud scavenging, whereby the wet aerosol particles are collected by falling raindrops (Barrie & Schemenauer, 1986). Wet deposition (via rain and other types of precipitation) is most efficient at removing soluble form of heavy metals from the air. A great percentage of metals fall through the rain at the place of their production (Nurnberg et al., 1984).

For example, dry deposition can remove gaseous mercury forms (particulate forms), while wet deposition is responsible for removing divalent mercury form (soluble form) (Nurnberg et al., 1984).

1.1.2.1.3 Occult deposition

Generally, this route is considered as one of the wet deposition processes. In occult deposition, wetted particles (fog and mist) are deposited by impaction or turbulent transfer (Barrie & Schemenauer, 1986).

However, heavy metals are principally dispersed by atmospheric transportation (Dobrovolsky, 1980), but it is necessary to consider that most atmospherically dispersed heavy metals enter the soil surface and the aboveground plant tissues very rapidly. In the soil, they are sorbed by the absorption complex and redistributed through the soil profile. A proportion of heavy metals is taken up by plants and leached by surface and subterranean runoff. The need to control atmospheric contamination of soils by heavy metals and take preventive and remediation measures is dictated by the serious ecological consequences of this contamination, as observed in different regions of the world (Galiulin et al., 2002).

2. Urgency of monitoring air contamination

The degree and extent of environmental changes over the last decades has given a new urgency and relevance to the detection and understanding of environmental change, due to human activities, which have altered global biogeochemical cycling of heavy metals and other pollutants. Approximately 5 million chemicals are presently known and 80,000 in use; 500 - 1,000 are added per year resulting in a progressive increase in the flux of bioavailable chemical forms to the atmosphere. Therefore air pollution has been one of the major threats to human health and the environment since the last century (Obbard et al., 2005; Batzias & Siontorou, 2006; Dmuchowski & Bytnerowicz, 2009).

2.1 Approaches for monitoring air contamination

There are two conceptual approaches for collecting samples relevant to air and atmospheric deposition related pollution studies. The first approach involves the direct collection of airborne particulate matter, precipitation and total deposit whereas the second approach uses air pollution biomonitors. The first approach is aimed at quantitative surveys at local, short-range, medium-range or global transport of pollutants, including health-related studies when collecting size fractionated airborne particulate matter. It requires continuous sampling on a long-term basis at a large number of sites, in order to ensure the temporal and spatial representativeness of measurements. The application of such direct measurements on a large scale is extremely costly and person-power intensive. Furthermore, it is not possible, due to logistic problems, to install instrumental equipment at all needed locations. Therefore, the second approach is considered as a non-expensive, yet reliable means of air quality status assessment in a country or a region. Certain types of biological organisms provide a measure of integrated exposure over a certain amount of time and enrich the substance to be determined so that the analytical accessibility is improved and the measurement uncertainty reduced. Sampling is relatively simple (even in remote areas and areas with difficult access) because of no expensive technical equipment is needed (Bagla et al., 2009). Furthermore, the sample treatment and analysis steps in the laboratory are facilitated (Wolterbeek, 2002) Also, this method avoids the need for deploying large numbers of precipitation collectors with an associated long-term program of routine sample collection and analysis (Harmens et al., 2008). Therefore, a much higher sampling density can be achieved than with conventional precipitation analysis and increases the possibility of monitoring many sites simultaneously (Wannaz et al., 2006). Also Bioindicators may be very useful due to their high sensitivity towards a broad spectrum of substances or because of their tolerance to high levels of a substance, accumulated in their tissues over an extended period of time or to integrate its influence in an area of known and relevant size (De Temmerman, 2004). Biomonitoring data are being compared with precipitation, particulate matter fractionation, and speciation data and also with medical statistics to evaluate the correlation between the amount of pollutants in the atmosphere and the human reaction to it (Szczepaniak & Biziuk, 2003).

2.2 Definition of biomonitoring and related terms

The term bioindicator/biomonitor is used to refer to an organism, or a part of it, that depicts the occurrence of pollutants on the basis of specific symptoms, reactions, morphological changes or concentrations (Markert et al., 1997). There is considerable variation in the use of the terms bioindicator and biomonitor: A bioindicator is an organism (or part of an organism or a community of organisms) that contains information on the quality of the environment (or a part of the environment). A biomonitor, on the other hand, is an organism (or part of an organism or a community of organisms) that contains information on the quantitative aspects of the quality of the environment. The clear differentiation between bioindication and biomonitoring using the qualitative/quantitative approach makes it comparable to instrumental measuring systems (Markert, 2007). Therefore biomonitoring, in the general sense, may be defined as the use of organisms and biomaterials to obtain information on certain characteristics of the biosphere. With proper selection of organisms, the general advantage of the biomonitoring approach is related primarily to the permanent and common occurrence of the organism in the field, even in remote areas, the ease of sampling, and the absence of any necessary expensive technical equipment (Wolterbeek, 2002).

2.3 Classification of biomonitors

Organisms can be classified according to the way in which the reaction is manifested: (1) reaction indicators, which have a sensitive reaction to air pollutants and which are used especially in studying the effects of pollutants on species composition, and on physiological and ecological functioning, and (2) accumulation indicators that readily accumulate a range of pollutants and are therefore used especially when monitoring the amount of pollutants and their distribution (Markert et al., 1997). Bioaccumulation monitoring methods can be divided into two groups: (Fig 2.)

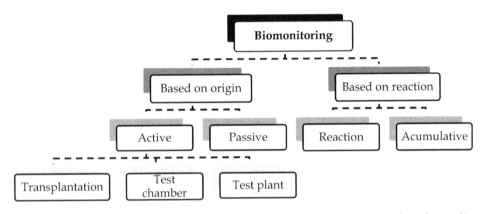

Fig. 2. General scheme of heavy metal biomonitoring using different materials and sampling techniques.

2.3.1 Active biomonitoring

Active biomonitoring includes the exposure of well-defined species under controlled conditions and can be divided into transplantation, test plant, and test chamber methods (C˘ eburnis & Valiulis, 1999). In the transplantation procedure, suitable organisms (mainly mosses and lichens) are transplanted from unpolluted areas to the polluted site under consideration. The exposure time thus is well defined, but the change in uptake efficiency due to climatic change is usually applied to testing either synergistic or single species effects of pollutants on sensitive biomonitors. Transplant techniques appear to be useful, in particular at relatively high pollutant levels. One distinct advantage, compared to the use of indigenous species, is that of well-defined exposure time, but the reproducibility of this technique appears not to be very satisfactory for parameters such as air concentration and deposition rate (C˘ eburnis and Valiulis, 1999).

2.3.2 Passive biomonitoring

Passive biomonitoring refers to the observation or chemical analysis of indigenous plants. In general, the passive biomonitoring method has one major disadvantage; all processes and all sources act at the same time and there is no possibility of separating them and looking for a particular one (C ˘ eburnis and Valiulis, 1999). Natural variabilities in ambient macro and microclimate conditions, such as acidity, temperature, humidity, light, and altitude (Gerdol et al., 2002), or ambient elemental (nutritional) occurrences may cause the biomonitors to exhibit variable behavior. Part of this variance is local (Wolterbeek & Verburg, 2002), but it may be clear that this variable behavior becomes a problem when it seriously affects the biomonitors in its accumulative exposure (Wolterbeek, 2002).

2.4 Principles of bioaccumulation monitoring

Biomonitoring is regarded as a means to assess trace element concentrations in aerosols and deposition. This implies that the monitor should concentrate the element of interest and quantitatively reflect its ambient conditions (Wolterbeek, 2002). In general, a good accumulation indicator of air pollutants should:

1. Accumulate pollutants from the air in the same way and to the same degree under different conditions (Wittig, 1993; Conti & Cecchetti, 2001).
2. The pollutants should be easily measured and the measurements should provide information about the level of pollutant deposition.
3. It should also indicate the risk limits caused by increasing levels of pollutants.
4. The species of organism used should be common enough and be available for collection throughout the year in the same area.
5. Its use should be based on standard sampling and analysis methods.
6. In order to determine the state of the ecosystem in relation to the pollutant under study, the state of the ecosystem in the background area should also be known (Seaward 1995).

The background level is usually considered to be the "natural" level at which emissions have as small an effect as possible (Conti & Cecchetti, 2001). The background level of different pollutants varies between plant species.

2.5 Type of biomonitors

As plants are immobile and more sensitive in terms of physiological reaction to the most prevalent air pollutants than humans and animals, they better reflect local conditions (Nali and Lorenzini 2007). For these reasons, plants are the most common used bioindicators in air quality biomonitoring studies (Balasooriya et al. 2009). Types of plant and their different parts have been applied in trace element air monitoring programs, such as lichens, mosses, ferns, grass, tree bark, tree rings, tree leaves and pine needles (Szczepaniak & Biziuk, 2003; Morselli et al., 2004). For all biomonitors used, the mechanisms of trace element uptake and retention are still not sufficiently known. For grass, tree rings, and ferns, substantial element contributions from other sources than atmospheric, such as the soil or the tree bole, have to be taken into account (Szczepaniak & Biziuk, 2003). Rossini Oliva and Mingorance (2006) reported that the accumulation pattern in the different parts of *Pinus pinea L.* and *Nerium oleander L.* was the following: wood < bark ≤ leaves, because of metals are taken up by wood stem from the soil and soil water, while the outer part (leaves and bark) intercept metals also by deposition from the atmosphere (Rossini Oliva & Mingorance, 2006).

It is important to note that a unique species that can be a suitable bioindicator for biomonitoring of toxic metal pollution all over the world has not been found yet. But lower and higher plants use as suitable biomonitors world-wide.

Plants accumulate metals due to many factors, such as element availability, the characteristics of the plants (such as species, age), state of health, and type of reproduction, temperature, available moisture, substratum characteristics, etc (Conti & Cecchetti, 2001)

2.5.1 Biomonitoring by lower plants

Lower plants are resistant to many substances which are highly toxic for other plants species. They are able to survive in such diverse and often extreme environment; these sedentary organisms possess an equally diverse set of physiological adaptations (Cenci et al., 2003; Fernandez et al., 2006; Dragovič & Mihailovič, 2009). Of all biological species used in biomonitoring, lichens and mosses have the most common occurrence. The morphology of lichens and mosses does not vary with seasons; thus accumulation can occur throughout the year. Lichens and mosses usually have considerable longevity, which led to their use as long-term integrators of atmospheric deposition (Szczepaniak & Biziuk, 2003).

2.5.1.1 Mosses as biomonitor

Botanically, mosses are bryophytes[1]. About 15,000 species of mosses are known world-wide (Richardson, 1981), most of which are abundantly distributed amongst the different ecosystems, growing on a variety of substrates (Onianwa, 2000). Two Swedish scientists Åke Rühling and Germund Tyler (1960) have discovered that mosses are good bioindicator of heavy metal pollution in the atmosphere, after this successful discovery many European countries have used mosses in national and multinational surveys of atmospheric-metal deposition. Mosses are cryptogams that thrive in a humid climate. Ectohydric mosses have been used as biomonitors – in most cases terricolous bryophytes. They possess many properties that make them suitable for monitoring air pollutants (Onianwa, 2000). These species obtain nutrients from wet and dry deposition because of they do not have real roots so there is no uptake of mineral substrates from other sources than the atmosphere. Nutrient uptake from the atmosphere is promoted by their weakly developed cuticle (so metal ions easily penetrate the cell wall), large surface to weight ratio, and their habit of growing in groups. Other suitable properties include a slow growth rate, small size and easy to handle, undeveloped vascular bundles (so transport of minerals between segments is limited), minimal morphological changes during the mosses' lifetime, perenniality, wide distribution (even in industrial and urban areas), an ability to survive in highly polluted environment, ease of sampling and the possibility to determine concentrations in the annual growth segments (Poikolainen, 2004; Dragovič and Mihailovič, 2009).

Air pollutants are deposited on mosses in aqueous solution, in gaseous form or attached to particles. The accumulation of pollutants in mosses occurs through a number of different mechanisms:

- As layers of particles or entrapment on the surface of the cells.
- Incorporation into the outer walls of the cells through ion exchange processes
- Metabolically controlled passage into the cells (Poikolaonen, 2004).

The attachment of particles is affected e.g., by the size of the particles and the surface structure of the mosses.

---

[1]Contains Mosses and Liverworts

Ion exchange is a fast physiological-chemical process that is affected e.g., by the number and type of free cation exchange sites, the age of the cells and their reaction to desiccation, growing conditions, temperature, precipitation pH, composition of the pollutants and leaching (Brown & Brûmelis 1996). In the ion exchange process, cations and anions become attached to functional organic groups in the cell walls among other things through chelation (Rao, 1984).

The cell wall has a high polyuronic acid content which makes moss a very good natural ion exchanger. The cell walls of bryophytes possess many negatively charged anionic sites to which cations are bound in exchangeable form. Studies of electron microscope have shown that the sorbed metal may be held either in the extracellular region outside of the cytoplasm, bound to the cell wall, and due to the highly reduced presence or absence of cuticle in the moss, ions have a direct access to the cell wall, mosses surfaces and rhizoids do not perform any active heavy metal ion discrimination (Shakya et al., 2008; Reimann et al., 2006; Onianwa, 2000).

Results of previous studies have shown that the degree of metal uptake efficiency retention proved to decrease in the order Cu > Pb > Ni > Co > Cd > Zn, Mn. Lead is very strongly fixed in the moss, and for this reason the correlation between concentration in moss and bulk deposition is particularly high (Szczepaniak & Biziuk, 2003; Rosman, et al., 1998).

Mosses have been used for monitoring atmospheric heavy metal levels in various forms, these include indigenous naturally-occurring epiphytic forms, moss transplants, moss-bags, and Sphagnum mosses. The majority of investigations have utilised epiphytic mosses growing naturally at a given location. Such mosses may be found growing on rocks, tree barks, walls or forest floors, and may be of the acrocarpous or pleurocarpous types. Many species of mosses have been so used, but perhaps the most commonly reported in the literature are the species *Hypnum cupressiforme*, *Hylocomium splendens*, and *Pleurozium schreberi*, particularly in parts of Europe where they are largely abundant (Onianwa, 2000). The comparisons between indigenous and transplanted mosses are summarized in table 2.

Factors affecting on uptake efficiency by mosses

The chemical composition of deposition has a large effect on the accumulation of pollutants, because the uptake efficiency of mosses for individual elements varies considerably (Berg et al., 1995). A high proportion of the pollutant load accumulates in mosses through wet deposition. The amount, duration and intensity of precipitation affect accumulation and leaching (Berg et al. 1995). Ross (1990) and Berg et al (1995) found the best correlation between the concentrations in mosses and in wet deposition for elements such as Pb, Cd, Co and Cu that have high uptake efficiency from wet deposition. The contribution of dry deposition increases on moving from humid to arid climates (Couto et al., 2004). Uptake efficiency is also affected by competition for free cation exchange sites; for instance, the presence of sea salts and acidic deposition has been found to have an effect on the absorbtion of metals by mosses (Gjengedal & Steinnes, 1990). The types of vegetation and soil dust have also been reported to cause regional differences in uptake efficiency (Szczepaniak & Biziuk, 2003). Other factors affecting the concentrations include stand throughfall and leaching from vegetation layers located above the mosses (Steinnes, 1993), the nutrient status of the site (Pakarinen & Rinne 1979; Økland et al., 1999) and snowmelt water (Ford et al. 1995). The altitude may also have an effect (Gerdol et al. 2002), due e.g., to changes in the amount of precipitation, dust or biomass production. The sampling and measuring methods employed can also have a considerable influence on the analytical results in biomonitoring studies (Steinnes et al. 1993; Wolterbeek & Bode 1999).

Indigenous	Transplants
Results of pollution patterns can be obtained within a few days.	A survey period of a year is required to allow for effects of seasonal variation.
Results demonstrate pollution in previous years.	Results illustrate pollution only over the sampling period.
Accumulation levels are usually above detectable levels due to longer exposure time.	Concentrations of accumulated levels may be undetectable over shorter sampling period.
Minimal supervision and risk from vandalism.	Potential risk from vandalism.
Costs acquired from transport to sites and chemical analyses.	Additional costs from materials, increased transport and more sampler preparation.
Potential shortage of indigenous samples.	Density of sampling sites, samplers and their position under the control of investigators.
Pollution deposition rates difficult to estimate.	Deposition rates calculated from controlled exposure time.
Metal concentrations reflect influences from other factors such as age of plant, metal content of substrate and local contamination.	Pollutant concentrations in plants can be more directly related to airborne pollution.
Plants may be stressed or undergo morphological/physiological changes, which affect uptake, by long-term exposure to certain pollutants.	

Table 2. Comparisons of indigenous samplers and transplants in heavy metal deposition monitoring (from Gailey and Lloyd, 1993).

It must be noted that the high concentrations of atmospheric pollutants such as sulphur dioxide and heavy metal particulates are harmful to the full development of mosses, and mosses do not thrive well around locations of such very high pollution levels (Onianwa, 2000).

2.5.1.2 Lichens as biomonitor

Lichens are the most studied biomonitors of air quality, since in 1866 a study was published on use of epiphytic lichens as bioindicators. They have been defined as "permanent control systems" for air pollution assessment (Conti & Cecchetti, 2001). Because of their high sensitivity toward specific pollutants and ability to store contaminants in their biological tissues, lichens are defined as bioindicators and/or bioaccumulators, respectively. Lichen form conspicuous gray, green, orange, or red patches on trees or rockes (Blasco et al., 2008).

Lichens are perennial cryptograms. They live on different types of substrate, usually on dry or nutrient-poor sites in boreal and sub-arctic regions (Nash 1996). The lichen species best suited as biomonitors are foliose[2] and fruticose[3] epiphytic lichens. Lichens consist of a symbiotic association of two organisms: the fungal component is usually an Ascomycetes

---

2. Some lichens have the aspect of leaves
3. Some lichens adopt shrubby forms

fungus (mycobiont), and the algal component (photobiont), a green alga (Chlorophyceae) and/or blue-green alga (Cyanobacteriae). The fungal component is responsible for taking up water and minerals, and the algal component, which grows amidst the fungal mycelia, for photosynthesis. Most lichen species obtain their nutrients from wet and dry deposition (Martin & Coughtrey 1982; Garty 1993). The lichen surface, structure, and roughness facilitate the interception and retention particles (Szczepaniak & Biziuk, 2003). They possess many of the same properties as mosses that make them suitable for monitoring purposes: the cuticle and vascular bundles are weakly developed, (means that the different pollutants are absorbed over the entire tallus surface of the organism), they do not have any real roots, they are slow-growing and long-lived, and they have an extremely broad distribution (Wolterbeek et al., 2003).

In general, three mechanisms have been reported with regard to the absorption of metals in lichens:

• Intercellular absorption through an exchange process
• Intercellular accumulation
• Entrapment of particles that contain metals (Szczepaniak & Biziuk, 2003).

Lichens may be used as bioindicators and/or biomonitors in two different ways: (1) By mapping all species present in a specific area; entitled "mapping lichen diversity" that consists recording changes in arboreal species diversity in fixed plots over time, either through the use of grid or transects samples, using a standardized method develop by lichenologists and (2) Through the individual sampling of lichen species and measurement of the pollutants that accumulate in the thallus, or by transplanting lichens from an uncontaminated area to a contaminated one and then measuring the morphological changes in the lichen thallus and/or evaluating the physiological parameters and/or evaluating the bioaccumulation of the pollutants (Conti & Cecchetti, 2001).

It was found that concentrations of Pb, Fe, Cu, Cr and Zn in *Hypogymnia physodes* correlated strongly with annual average atmospheric deposition (Sloof, 1995).

In the context of pollution monitoring which uses naturally occurring lichens at monitoring sites as well as lichen transplants, changes in lichen morphology, cytology, metabolism and physiological processes, all provide environmental insight. Normal *Hypogymnia physodes* (L.) Nyl., for example, has a smooth, grey thallus with large lobes but when exposed to air pollution, the surface becomes cracked and the coloration brown or black (Richardson & Nieboer, 1981)

Factors affecting on uptake efficiency by lichen

There are a considerable number of factors, associated with the site where lichens are growing, which may change the concentrations of pollutants in lichens (Brown, 1991; Garty, 2000). These factors are, in most cases, the same as those affecting mosses including quality of the deposition (form of occurrence, composition, pH), climate (composition of precipitation, temperature, wind, drought, length of the growing period) and local environmental factors (vegetation, quality of the substrate, stand throughfall and stemflow, dust derived from soil, altitude of area). On the other hand, throughfall and stemflow, which vary according to the type of canopy cover, have a greater effect on epiphytic lichens than on terricolous mosses (Poikolainen, 2004; Rasmussen, 1978).

Most species are especially sensitive to SO_2, nitrogen and fluoride compounds and to ozone. These compounds affect the condition of lichens and thus reduce the capacity of lichens to accumulate and absorb elements from the atmosphere (Poikolainen, 2004).

Air pollutants have a different effect on the fungal and on the algal partner. The algal partner has been reported to react more sensitively e.g., to acidic deposition and heavy metals, and to show varying accumulation of metals depending on the acidity of precipitation. Sporadic desiccation of lichens may also have an effect on the accumulation and absorption of elements (Poikolainen, 2004). After a dry period, rainfall may result in appreciable washing off particles and the exchange of cations bound on negatively charged exchange sites on the cell walls and plasma membranes of the cells (Bargagli, 1998).

Results of different studies about heavy metal concentrations in various lichen species are summarized in table 3. There are clear differences in the accumulation of elements between different lichen species (even in the same studies) as a result of morphological and physiological differences.

Lichens (Bowen, 1979)	Pseudevernia furfuracea (Adamo, 2003)	Foliose (e.g. Parmelia, Hypogymnia) (Adamo, 2003)	Evernia prunastri (Conti, 2004)	Dirinaria picta (NG et al., 2006)	C.rangiformis (Cayir et al., 2007)	Pyxine cocoes (Rajesh et al., 2011)	Phaeophyscia hispidula (Rajesh et al., 2011)
Cd -	0.46	(<0.1-0.3)	(0.05-0.09)	(0.14-0.28)	(0.14-0.69)	(0.9-6.3)	(0.8-6.8)
Cr (0.6-7.3)	2.23	(1-4)	(1.04-2.81)	-	(2.24-13.0)	(0.8-26.2)	(3.4-35.2)
Cu (9-24)	5.42	(4-10)	(1.94-4.45)	(11.75-45.13)	(1.06-5.29)	-	-
Pb (1-78)	23	(1-8)	(1.05-3.62)	(2.83-16.59)	(1.35-33.8)	(0.1-13.3)	(4.4-11.7)
Zn (20-60)	99	(20-90)	(20.3-53.2)	(44.17-83.15)	(9.15-47.6)	(57.3-194.4)	(103.1-214.6)

Table 3. Heavy metal concentrations in lichen species from different studies (mg/kg), (ranges in brackets).

2.5.1.3 Moss to lichens comparison

As above mentioned mosses and lichens do not have root systems like higher plants therefore their contaminant content depend on surface absorption. These organisms have been shown to concentrate particulates and dissolve chemical species from dry and wet deposition. Differences in substrata result in differences in lichen metal content. For mosses, relatively high contribution of crustal elements such as Al, Sc, La and lanthanides is observed. The increase in cation exchange capacity from moss apex to base is apart of its natural balance of elements, which in turn is affected by the proximity of the soil.

Lichen identification and collecting turns out to be very complicated, whereas for mosses it is much easier. The annual growth increment is easier to distinguish for mosses than for lichens; therefore they are considered superior to lichens, if any time resolution in the measurement is required. Older parts of lichens carry fruiting bodies rich in metals. Mosses build carpets during a period of 3–5 years, and their metal content is generally considered to reflect the atmospheric deposition during that period (Szczepaniak & Biziuk, 2003). Loppi and Bonini showed that lichens and mosses can be used indifferently as accumulators of As, Cd, Cu, Mo, and S. Differences in concentrations between lichens and mosses were statistically significant ($P<0.05$) for Al, B, Fe, Hg, Pb, Sb and Zn, with mosses retaining higher values than lichens except for Hg and Zn. The elements found in higher concentrations in mosses were associated with particulate matter (Szczepaniak & Biziuk, 2003). Bargagli et al. found higher concentrations of lithophile elements (Al, Cr, Fe, Mn, Ni, and Ti) in moss and atmophile elements (Hg, Cd, Pb, Cu, V, and Zn) in lichen (Bargagli et al., 2002). According to studies of Kansanen and Venetvaara (1991) mosses and lichens as the most effective indicators for low and moderate level of metal deposition in polluted areas.

2.5.2 Biomonitoring by higher plants

However mosses and lichens most frequently used for monitoring metal pollution, but these lower plants are characterized by irregular and patchy distribution and their sampling should be done by specialists who can differentiate between similar-looking species. These limitations become more pronounced in industrial and densely populated areas, where sever anthropogenic pressure may cause scarcity or even lack of indicator species at some sampling points. For instance lichens are characterized by slow regeneration and relatively weak tolerance to the complex influence of mycophytotoxic pollutants, Therefore intensive sampling may lead to their reduced availability and even disappearance (Berlizov et al., 2007).

The use of higher plants, especially different parts of trees (leaves and barks), for air monitoring purposes is becoming more and more widespread. The main advantages are greater availability of the biological material, simplicity of species identification, sampling and treatment, harmless sampling and ubiquity of some genera, which makes it possible to cover large areas. Higher plants also exhibit greater tolerance to environmental changes which is especially important for monitoring areas with elevated anthropogenic influence (Berlizov et al., 2007), therefore higher plants have appeal as indicators in air pollution monitoring in highly polluted areas where lichens and mosses are often absent.

Higher plants not only intercept pollutants from atmospheric deposition but also accumulate aerial metals from the soil. Aerial heavy metal depositions are taken up from the soil by plants via their root system and translocated them to other parts of the plant (Mulgrew & Williams, 2000). An otherwise in the industrial and urban areas, higher plants can give better quantifications for pollutant concentrations and atmospheric deposition than non-biological samples (Markert, 1993).

Some plant species are sensitive to single pollutants or to mixtures of pollutants. Those species or cultivars are likely to be used in order to monitor the effects of air pollutants as bioindicator plants (DeTemmerman et al., 2005).

In general, it can be assumed two separated groups of higher plants for biomonitoring purposes including herbs/grasses and trees/shrubs on both groups, aboveground plant tissues (leaf and bark) contribute in airborne heavy metal accumulation.

2.5.2.1 Herbs and grasses as biomonitor

Kovács (1992a) recommended the use of ruderal plants as bioaccumulative indicators due to their ability to accumulate metals in high quantities without visible injury (Kovács, 1992a).

Some plant species may be more efficient in retaining atmospheric metal particles than others. A measure of this efficiency can be resolved by calculating air accumulation factors (AAF) according to the following equation:

$$AAF \ (m^3.g^{-1}) = PAc \ (mg. \ g^{-1} \ dry \ weight)/CA \ (mg. \ m^{-3}) \qquad (1)$$

PAc = atmospheric contribution of the metal in plants and CA = concentration of the metal in the atmosphere (Mulgrew & Williams, 2000).

2.5.2.2 Trees and shrubs as biomonitor

Both coniferous and deciduous trees can be used in the detection of aerial heavy metal pollution, but Coniferous trees indicate pollution over a longer time period such as, *Pinus eldarica, Cupressus arizonica* (Ataabadi et al., 2010a), *Pinus brutia* (Baslar et al., 2009), *Cupressus semervirens* (El-Hasan et al., 2002). Broad-leaved tree species regarded as sensitive to metal contamination include *Betula pendula, Fraxinus excelsior, Sorbus aucuparia, Tilia cordata* and *Malus domestica* (Mulgrew & Williams, 2000). Numerous bioaccumulative indicators exist; some tree examples include *Ailanthus glandulosa, Celtis occidentalis, Salix alba, Tilia tomentosa, Sambucus nigra, Quercus robur, Fagus silvatica* (Mulgrew & Williams, 2000) and *Elaeagnus angustifolia* (Akosy and Sahin, 1999). Also results of studies indicate that *Robinia pseudoacacia* (black locust tree) is appropriate species because of this tree is genetically homogeneous, easily identifiable and ubiquitously distributed (Kovács, 1992b).

However, there are limited studies about biomonitoring by shrub species, but it could numbered some shrub species such as *Nerium oleander* (Ataabadi et al., 2010a; Rossini Oliva & Mingorance, 2006), *Lantana Camara* (Fernandez Espinoza & Rossini Oliva, 2005) *Ligustrum vulgare, Photinia serrulata, Berberis vulgaris* and *Thuja orientalis* (Ataabadi et al., 2010a,b; Ataabadi et al., 2011).

For comparative studies, it is important that sampling is undertaken at the same time of the year to reduce variability. Chemical composition of foliage varies with season and rainfall (Taylor et al., 1990). For most deciduous species, suitable time is period of year when metal content in leaves will be highest. For instance standard sampling of heavy metal accumulation in *Populus nigra* in central Europe is carried out in August.

Metal content will vary depending on which part of the plant is sampled. The extent of accumulation in different plant parts will vary with species and the nature of the element.

It should be noted that using transplantation exercises are not common for higher plants (Mulgrew & Williams, 2000).

2.5.2.2.1 Plant leaves as biomonitor

Leaves of higher plants have been used for heavy metals biomonitoring since 1950s (Al-Shayeb eta l., 1995). The use of leaves as bioindicators of environmental pollution has been studied, more and more, to assess their suitability, to assess effect of a specific pollution source, to differentiate between background (unpolluted) and polluted sites and to monitor or assess the level of pollution in an area (Turan et al., 2011).

Rossini Oliva & Mingorance (2006) and Ataabadi et al (2010a), reported that pine needles can be considered suitable biomonitor for atmospheric heavy metal contamination (e.g. Fe, Al, Pb).

Factors that affect efficiency of heavy metal accumulation on leaf surface

Particle (containing heavy metals) deposition on leaf surfaces may be affected by two different factors including plant-dependent factors such as morphological and structural properties of leaves contain orientation and size (Mulgrew & Williams, 2000) of leaves, cuticle thickness, cork existence, roughness, existence of surface waxy layer, specific leaf area (SLA), stomatal density (SD) and stomatal pore surface (SPS) (Ataabadi et al., 2011; Rossini Oliva & Mingorance, 2006; Balasooriya et al. 2008) and plant-free factors such as particle size and wind velocity (Mulgrew & Williams, 2000). Also the accumulation of heavy metals by higher plants depends on the binding and solubility of particles deposited on leaf surfaces (Mankovska et al., 2004).

The deposited particles may be washed by rain into the soil, resuspended or retained on plant foliage. The degree of retention is influenced by weather conditions, nature of pollutant, plant surface characteristics and particle size (Harrison and Chirgawi, 1989).

Leaves of evergreen species are considered to be better traps because of higher accumulation on a longer period of time (Turan et al., 2011).

2.5.2.2.2 Plant bark as biomonitor

The physiological function of bark is to protect the tree from mechanical injury, damaging agents and excessive evaporation. Bark quality varies considerably in different tree species and at different stages in the lifetime of a tree species. The outer bark of trees consists of the inner layer (phloem), the cork-forming layer (phellogen), and the outer layer (rhytidom or phellem) composed of dead cork cells (Prance et al., 1993). This dead cork layer has usually been employed in biomonitor studies. The chemical composition of the bark is specific to each tree species. For instance, the pH, electrical conductivity and ash content of the bark of coniferous trees are usually lower than those of broad-leaved species (Barkman, 1958). When bark is exposed to air pollutants either directly from the atmosphere or from the rainwater running down the stem, the chemical composition of the surface layers of the bark changes, such changes can be utilised in investigating the extent of the area subjected to air pollutants. In this respect tree bark is a good bioindicator because it remains in place for an extended period of time, it is easily accessible and sampling does not damage the tree (Berlizov eta al., 2007). Retention of suspended particles is promoted by a moist, porous, rough, or electrically charged surface, making bark a highly effective collector (Panichev et al., 2004). Therefore bark has been widely employed as a passive monitor for airborne metal contaminations. The accumulation of atmospheric pollutants in bark is purely a physiological-chemical process. The pollutants either accumulate passively on the bark surface or become absorbed through ion exchange processes in the outer parts of the dead cork layer (Poikolainen, 2004). Although a number of air pollution biomonitoring studies have been performed using bark of different tree species include oak, elm, willow, poplar, pine, olive, cedar, eucalyptus etc. (Berlizov et al., 2007).

Metal accumulation in bark and plant foliage in urban and industrial areas can be considerable, with the greatest amount of the heavy metal burden located in the bark (Ce^burnis & Steinnes, 2000; Watmough & Hutchinson, 2003).

Factors that affect efficiency of heavy metal accumulation on bark surface

Factors, in addition to atmospheric pollutants, that affect the chemical composition of tree bark are mainly the same as those for mosses and lichen, although the chemical reactions that occur in bark are somewhat different because bark is a non-living plant material. The concentrations in bark are mainly affected by bark quality, stand throughfall and stem flow. The concentrations are highest in the surface layers of the outer bark, and decrease rapidly on moving towards the inner layers. Many different factors have an effect on the collecting of heavy metals in bark surface, such as heavy metal quantities in air, physiological and chemical properties of the bark, through fall, soil factors, contamination of other plants, climate factors, etc. A coarse and rough surface more readily accumulates atmospheric pollutants than a smooth surface (Poikolainen, 2004), As Barnes et al., (1976) showed that rough barks accumulate metals more than smooth barks. Other factors include the bark texture and thickness (Poikolainen, 2004), the presence of epiphytic organisms, the time of exposure to the atmosphere and the depth of sampling (Bellis et al., 2001). Also bark acidity has an effect on the concentrations of some heavy metals. For instance, Bates and Brown (1981) found a clear negative correlation between bark pH and the Fe concentration in a study on the occurrence of epiphytic lichens on oak and ash. They concluded that this is due to the increased mobility of Fe with decreasing bark pH. There is no significant migration of elements from the bark surface through the cork tissue into the underlying wood, or vice versa. The migration of heavy metals from the soil via the roots into the bark as it is being formed is also usually insignificant. On the other hand, heavy metals and other compounds may be carried by the wind from the soil to the bark surface (Poikolainen, 2004).

The study carried out by Szopa et al (1973) on lead concentrations along highways in the US indicated that the lead concentration in bark reacts rapidly to marked changes in lead concentrations in the atmosphere (Poikolainen, 2004).

2.6 Analytical method

Finally, the choice of analytical method for heavy metal detection in all plant materials depends on the purpose of the respective survey. Some analytical methods are non-destructive (e.g. Neutron Activation Analysis: N.A.A.) and are useful for repetitive surveys such as baseline studies. Samples can also be archived and used at a later date for additional analysis. Destructive techniques include atomic absorption spectrometry (AAS) and inductively coupled plasma (ICP) analysis (Mulgrew & Williams, 2000).

On consideration of the methods available for moss monitoring in Norway, Steinnes et al. (1993) concluded that ICP-ES works well for Fe, Zn, Pb and Cu, to a lesser extent for V and Ni and but is not satisfactory for Cr, Cd and As. ICP-MS analysis proves a good method for all of the above except As and Cd where less satisfactory results were observed (Steinnes, 1993).

2.7 Some atmospheric heavy metal contamination indices

Several integrated environmental indices reflecting the quality of environment and life have begun to appear in recent years. On the other hand, quantitative information can be obtained by calculating different indices to estimate air quality (Calvelo & Liberatore, 2004).

2.7.1 Index of Atmospheric Purity (IAP)

DeSloover(1964) and DeSloover and LeBlanc (1968) suggested the IAP method for mapping air pollution on the basis of epiphytic lichen and bryophyte sensitivity for mapping aerial pollution (Moore, 1974). This method is based on the fact that the epiphytic lichen and mosses diversity is impaired by air pollution and environmental stress. Therefore the frequency of occurrence of species on a defined portion of a tree trunk is used as a parameter to estimate the degree of environmental stress. This index gives an evaluation of the level of atmospheric pollution, which is based on the number, frequency, and tolerance of the epiphytic species present in an area.

$$IAP = \sum_{i=1}^{n}(Q_i \times f_i) \tag{2}$$

n, the number of epiphytic species per site; Q_i, the resistance factor or ecological index of each species, thus representing the sensibility of a species against pollutants; f_i, the frequency or coverage score of each species per site (Max. 10; $1 \le f \le 10$). The IAP-index was calculated separately for each sampling site (Blasco et al., 2008; Zechmeister & Hohenwallner, 2006).

The frequency method makes it possible to predict pollution levels with a certainty of over 97% (Conti & Cecchetti, 2001). After calculation of IAP, pollution intensity will be determined by IAP classification is shown in Table 4.

Level A	$0 \le IAP \le 12.5$	Very high level of pollution
Level B	$12.5 \le IAP \le 25$	High level of pollution
Level C	$25 \le IAP \le 37.5$	Moderate level of pollution
Level D	$37.5 \le IAP \le 50$	Low level of pollution
Level E	$IAP > 50$	Very low level of pollution

Table 4. Quality levels of index of atmospheric purity (IAP) (from Szczepaniak & Biziuk, 2003).

Quality levels may somewhat differ in various studies. As Dymytrova (2009) studied a total of 1730 trees and 272 sampling plots in different parts of Kyiv (Ukraine) including industrial areas, residential areas, roads and inner parks and investigated epiphytic bryophytes and lichens on isolated trees. Results showed the highest and lowest epiphytic richness in the inner parks and industrial area respectively. Consequently based on IAP, four zones with different air pollution were distinguished: highly polluted (0-8.5), moderately polluted (8.6-20.2), slightly polluted (20.3-31.9) and unpolluted (32-88.8). In these studies, heavy metal concentrations in epiphytic vegetation and quality levels of IAP are correlated.

2.7.2 Enrichment factor

One approach used to characterize airborne particulate matter in terms of chemical composition is to calculate so called 'enrichment factors' (EF), relating the concentration of an anthropogenic 'pollutant' element (X), such as Pb, to that of a crustal element (typically Al, Ti, Sc or Fe) in air, normalized to the ratio of these elements in the average continental crust:

$$EF = (X_{air}/Al_{air}) / (X_{crust}/Al_{crust}) \tag{3}$$

Enrichment factors close to unity thus indicate that windblown dusts are the dominant airborne source, whilst values in excess of unity indicate that analyte concentrations have been elevated as a result of anthropogenic inputs. The calculation assumes that the anthropogenic contribution of the normalizing element (i.e. Al) is insignificant (Zoller et al., 1974; Puxbaum, 1991).

Pacheco et al. (2002) evaluated olive-tree bark for the biological monitoring of airborne trace elements and found significant enrichment of Cu in olive bark relative to *Parmelia* spp. lichen was reported (Pacheo et al., 2002).

3. Conclusion

Application of lower and higher plants as biomonitors seams to be a good way to monitor airborne heavy metal contamination, but the choice of proper phytomonitor for environmental studies depends on many factors such as availability of the biological material, contamination extent, study scale and etc. However lichen and moss due to have superior ability to accumulate elements and indicate them without interference with soil are reported the best bioindicators for the atmospheric heavy metal contamination, but in the urban and industrial areas, using aboveground parts of indigenous higher plants is recommended. Simplicity of species identification, sampling and treatment and ubiquity of some genera makes it possible to cover large areas. For the comprehensive conclusion, both of lower and higher plants should be studied simultaneously in the specific area and manifested with respect to their limitations and advantages.

4. Acknowledgment

The authors would like to thank Dr. Ahmad Ali Foroughi, the chancellor of Islamic Azad University Khorasganan branch (Isfahan) and Research Center of this University.

5. References

Adamo, P.; Giordano, S. Vingiani, S. Cobianchi, R.C. & Violante, P. (2003). Trace Element Accumulation by Moss and Lichen Exposed in Bags in the City of Naples (Italy). *Environmental Pollution*, Vol.122, No.1, (2003), pp. 91–103, ISSN 0269-7491

Adriano, D. C. (1986). Trace Elements in the Terrestrial Environment. Springer, New York.

Akosy, A. (2008). Chicory (*Cichorium intybus* L.): A Possible Biomonitor of Metal Pollution. *Pakistan Journal of Botany*, Vol.40, No. 2, (1986), pp. 791-797, ISSN 0556-3321

Akosy, A.; & Sahin, U. (1999). *Elaeagnus angustifolia L.* as a Biomonitor of Heavy Metal Pollution. *Turkish Journal of Botany*, Vol.23, No.2, (1999), pp. 83-87, ISSN 1300-008X

Al-Shayeb, S.M.; Al-Rajhi, M. A. & Seaward, M.R.D. (1995).The Date Palm (Phoenix dactylifera L.) as a Biomonitor of Lead and Other Elements in Arid Environments. *Science of the Total Environment*, Vol.168, No.1, (1995), pp. 1-10, ISSN 0048-9697

Ataabadi, M.; Hoodaji, M. & Najafi, P. (2010a). Heavy Metal Biomonitoring by Plants Grown in an Industrial Area of Isfahanò Mobarakeh Steel Company. *Journal of Environmental Studies*, Vol.35, No. 53, (2010), pp. 83-92, ISSN 1025-8620

Ataabadi, M.; Hoodaji, M. & Najafi, P. (2010b). Evaluation of Airborne Heavy Metal Contamination by Plants Growing Under Industrial Emission. *Environmental engineering and Management Journal*, Vol.9, No. 7, (2010), pp. 903-908, ISSN 1582-9596

Ataabadi, M.; Hoodaji, M. & Najafi, P. (2011). Biomonitoring of Some Heavy Metal Contaminations from a Steel Plant by Above Ground Plants Tissue. *African Journal of Biotechnology*, Vol.10, No.20, (2011), pp. 4127-4132, ISSN 1684–5315

Avino, P.; Capannesi, G. & Rosada, A. (2008). Heavy Metal Determination in Atmospheric Particulate Matter by Instrumental Neutron Activation Analysis. *Microchemical Journal*, Vol.88, No.2, (2008), pp.97–106, ISSN 0026-265X

Bajpai, R.; Mishra, G.K. Mohabe, S. Upreti, D.K. & Nayaka, S. (2011). Determination of Atmospheric Heavy Metals Using Two Lichen Species in Katni and Rewa cities, India. *Journal of Environmental Biology*, Vol.32, No.2, (2011), pp. 195-199, ISSN 0254-8704

Balasooriya, B.L.W.K.; Samson, R. Mbikwa, F. & Vitharana, U.W.A. (2009). Biomonitoring of
    Urban Habitat Quality by Anatomical and Chemical Leaf Characteristics.
    Environmental and Experimental Botany, Vol.65, No. 2-3, (2009), pp. 386-394, ISSN
    0098-8472
Bargagli, R. (1998). *Trace elements in terrestrial plants: An Ecophysiological Approach to
    Biomonitoring and Biorecovery*, Springer, ISBN 3540645519, Berlin, New York
Bargagli, R.; Monaci, F. Borghini, F. Bravi, F. & Agnorelli, C. (2002). Mosses and Lichens as
    Biomonitors of Trace Metals. A Comparison Study on Hypnum cupressiforme and
    Parmelia caperata in a Former Mining District in Italy. *Environmental Pollution*,
    Vol.116, No.2, (2002), pp. 279–287, ISSN 0269-7491
Barnes, D.; Hammadah, M.A. & Ottaway, J.M. (1976).The Lead, Copper, and Zinc Contents
    of Tree Rings and Barks, a Measurement of Local Pollution. *Science of the Total
    Environment*, Vol.5, No.1, (1976), pp. 63– 77, ISSN 0048-9697
Barrie, L.A. & Schemenauer, R.S. (1986). Pollutant Wet Deposition Mechanisms in
    Precipitation and Fog Water. *Water, Air, & Soil pollution*, Vol.30, No. 1-2, (1986), pp.
    91-104, ISSN 0049-6979
Baslar, S.; Dogan, Y. Durkan, N & Bag, H. (2009). Biomonitoring of Zinc and Manganese in
    Bark of Turkish Red Pine of Western Anatolia. *Journal of Environmental Biology*,
    Vol.30, No. 5, (2009), pp. 831-834, ISSN 0254-8704
Bates, J.W. & Brown, D.H. (1981). Epiphyte Differentiation Between *Quercus petraea* and
    Fraxinus excelsior Trees in a Maritime Area of South West England. *Vegetatio*, Vol.48,
    No.1, (1981), pp. 61-70, ISSN 00423106
Batzias, F. A. & Siontorou, Ch. G. (2006). A knowledge - based Approach to Environmental
    Biomonitoring. *Environmental Monitoring and Assessment*, Vol.123, No.1-3, (2006),
    pp. 167-197, ISSN 0167-6369
Beijer, K.; & Jernelöv, A. (1986). Sources, Transport and Transformation of Metals in the
    Environment, In: *Handbook on the Toxicology of Metals*, L. Frindberg, G.F. Nordberg,
    V.B. Vouk, (Eds.), 68–74, Elsevier Scientific Publication, ISBN 0444904131,
    Amsterdam
Bellis, D.; Cox, A.J. Staton, I. McLeod, C.W. & Satake, K. (2001). Mapping Airborne Lead
    Contamination Near a Metals Smelter in Derbyshire, U.K.: Spatial Variation of Pb
    Concentration and 'Enrichment Factor' for Tree Bark. *Journal of Environmental
    Monitoring*, Vol.3, No.5, (2001), pp. 512–514, ISSN 1464-0325
Berg, T.; Røyset, O. & Steinnes, E. (1995). Moss (*Hylocomium splendens*) Used as Biomonitor
    of Atmospheric Trace Element Deposition: Estimation of Uptake Efficiencies.
    Atmospheric Environment, Vol.29, No.3, (1995), pp. 353-360, ISSN 1352-2310
Berlizov, A.N.; Blum, O.B. Filby, R.H. Malyuk, I.A. & Tryshyn, V.V. (2007). Testing
    Applicability of Black Poplar (Populus nigra L.) Bark to Heavy Metal Air Pollution
    Monitoring in Urban and Industrial Regions. *Science of the Total Environment*,Vol.
    372, No.2-3, (2007), pp. 693-706, ISSN 0048-9697
Blasco, M.; Domeno, C. & Nerin, C. (2008). Lichens Biomonitoring as Feasible Methodology
    to Assess Air Pollution in Natural Ecosystems: Combined study of quantitative
    PAHs Analyses and Lichen Biodiversity in the Pyrenees Mountains. *Analytical and
    Bioanalytical Chemistry*, Vol.391, No.3, (2008), pp. 759–771, ISSN 1618-2642
Bowen, H.J.M. (1979). *Environmental Chemistry of the Elements*, Academic Press, ISBN
    0121204502, New York

Brown, D.H. & Brûmelis, G. (1996). A Biomonitoring Method Using the Cellular Distribution of Metals in Moss. *Science of the Total Environment*, Vol.187, No.2, (1996) pp. 153-161, ISSN 0048-9697

Brown, D.H. (1991). Lichen Mineral Studies – Currently Clarified or Confused. *Symbiosis*, Vol.11, No.2-3, (1991), pp. 207-223, ISSN 0334-5114

C˘eburnis, D. & Valiulis, D. (1999). Investigation of Absolute Metal Uptake Efficiency from Precipitation in Moss. *Science of the Total Environment*, Vol.226, No.2-3, (1999), pp. 247–253, ISSN 0048-9697

Calvelo, S. & Liberatore, S. (2004). Applicability of In Situ or Transplanted Lichens for Assessment of Atmospheric Pollution in Patagonia, Argentina. *Journal of Atmospheric Chemistry*, Vol.49, No.1-3, (2004), pp. 199–210, ISSN 0167-7764

Cayir, A.; Coskun, M. & Coskun, M. (2007). Determination of Atmospheric Heavy Metal Pollution in Canakkale and Balikesir Provinces Using Lichen (Cladonia rangiformis) as a Bioindicator. *Bulletin of Environmental Contamination and Toxicology*, Vol.79, No.4, (2007), pp. 367–370, ISSN 0007-4861

Ce^burnis, D. & Steinnes, E. (2000). Conifer Needles as Biomonitors of Atmospheric Heavy Metal Deposition: Comparison with Mosses and Precipitation, Role of the Canopy. *Atmospheric Environment*, Vol.34, No.25, (2000), pp. 4265– 4271, ISSN 1352-2310

Cenci, R.M.; Sena, F. Bergonzoni, M. Simonazzi, N. Meglioli, E. Canovi, L. Locoro, G. & Trincherini, P. (2003). Use of Mosses and Soils for the Monitoring of Trace Elements in Three Landfills, Used as Urban Waste Disposal Sites, *Proceedings of Sardinia 2003 9th International Waste Management and Landfill Symposium*, Italy, October 6-10,2003

Conti, M.E. & Cecchetti, G. (2001). Biological Monitoring: Lichens as Bioindicators of Air Pollution Assessment--a Review. *Environmental Pollution*, Vol.114, No.3, (2001), pp. 471–492, ISSN 0269-7491

Conti, M.E.; Tudino, M. Stripeikis, J. Cecchetti, G. (2004). Heavy Metal Accumulation in the Lichen *Evernia prunastri* Transplanted at Urban, Rural and Industrial Sites in Central Italy. Journal of Atmospheric Chemistry, Vol.49, No.1-3, (2004), pp. 83-94, ISSN 0167-7764

Couto, J.A.; Fernández, J.A. Aboal, J.R. & Carballeira, A. (2004) Active Biomonitoring of Element Uptake with Terrestrial Mosses: a Comparison of Bulk and Dry Deposition. *Science of the Total Environment*, Vol.324, No.1-3, (2004), pp. 211-222, ISSN 0048-9697

De Temmerman, L.; Bell, J.N.B. Garrec, J.P. Klumpp, A. Krause, G.H.M. & Tonneijck, A.E.G. (2004). Biomonitoring of Air Pollution with Plants Considerations for the Future, In: *Urban Air Pollution, Bioindication and Environmental Awareness*, A. Klumpp, W. Ansel, G. Klumpp, (Eds.), 337-373, Cuvillier Verlag, ISBN 3865370780, Go¨ttingen

De Temmerman, L.; Nigel, J. Bell, B. Garrec, J.P. Klumpp, A. Krause, G. H. M. & Tonneijck, A. E. G. (2005). Biomonitoring of Air Pollutants with Plants, In: *International society of environmental Botanists*, Vol.11, No.2, Available form:
http://isebindia.com/05_08/05-04-1.html

Dmuchowski, W. & Bytnerowicz, A. (2009). Long-term (1992–2004) Record of Lead, Cadmium, and Zinc Air Contamination in Warsaw, Poland: Determination by Chemical Analysis of Moss Bags and Leaves of Crimean linden. *Environmental pollution*, Vol.157, No.12, (2009), pp. 3413-3421, ISSN 0269-7491

Dobrovolsky, V.V. (1980). Heavy Metals: Environmental Contamination and Global Geochemistry, In: *Heavy Metals in the Environment*, V.V. Dobrovolsky, (Ed.), 3-12, Moscow State University Publishing House, Moscow

Dragovič, S. & Mihailovič, N. (2009). Analysis of Mosses and Topsoils for Detecting Sources of Heavy Metal Pollution: Multivariate and Enrichment Factor Analysis. *Environmental Monitoring and Assessment*, Vol.157, No.1-4, (2009), pp. 383-390, ISSN 0167-6369

Duffus, J.H. (2002). "Heavy metal"- A Meaningless Term? *Pure and Applied Chemistry*, Vol. 74, No.5, (2002) pp. 793–807, ISSN 0033-4545

Dymytrova, L. (2009). Epiphytic lichens and bryophytes as indicators of air pollution in Kyiv city (Ukraine). *Folia Cryptogamica Estonica*, Vol.46, No.1, pp.33-44, ISSN 1406-2070

El-Hassan, T.; Al-Omari, H. Anwar, J. Al-Nasir, F. (2002). Cypress Tree (*Cupressus semervirens L.*) Bark as an Indicator for Heavy Metal Pollution in the Atmosphere of Amman City, Jordan. *Environment International*, Vol.28, No.6, (2002), pp. 513-519, ISSN 0160-4120

Fernandez Espinoza, A.J. & Rossini Oliva, S. (2005). The Composition and Relationships Between Trace Element Levels in Inhalable Atmospheric Particles (PM_{10}) and in Leaves of *Nerium oleander* L. and *Lantana camara* L. *Chemosphere*, Vol.62, No.10, (2005), pp. 1665–1672, ISSN 0045-6535

Fernandez, C. C.; Shevock, R. Glazer, N. A. & Thompson, J. N. (2006). Cryptic Species within the Cosmopolitan Desiccation-Tolerant Moss *Grimmia laevigata*. *Proceedings of the National Academy of Sciences of the United States of America*, Vol.103, No.3, (2006), pp. 637 – 642, ISSN 1091-6490

Gailey, F.A.Y. & Lloyd, O.L. (1993). Spatial and Temporal Patterns of Airborne Metal Pollution: The Value of Low Technology Sampling to an Environmental Epidemiology Study. *Science of the Total Environment*, **Vol.133**, No.3, (1993), pp. 201-219, ISSN 0048-9697

Galiulin, R.V.; Bashkin, V.N. Galiulina, R.A. & Kucharski, R. (2002). Airborne Soil Contamination by Heavy Metals in Russia and Poland, and its Remediation. *Land Contamination & Reclamation*, Vol.10, No.3, (2002), pp. 179-187, ISSN 0967-0513

Galloway, J.N.; Thornton, J.D. Norton, S.A. Volchok, H.L. & McClean, H.L. (1982). Trace Metals in Atmospheric Deposition: A Review and Assessment. *Atmospheric Environment*, Vol.16, No.7, (1982), pp. 1677-1700, ISSN 1352-2310

Garty, J. (1993). Lichens as Biomonitors for Heavy Metal Pollution, In: *Plants as Biomonitors Indicators for Heavy Metals in the Terrestrial Environment*, B. Markert, (Ed.), 193-263, VCH, ISBN 3527300015,Weinheim

Garty, J. (2000). Environment and Elemental Content of Lichens, In: *Trace elements – Their Distribution and Effects in the Environment*, B. Markert & K. Friese (Eds.), 33-86, Elsevier Science, ISBN 0444505326, Amsterdam

Gerdol, R.; Bragazza, L. & Marchesini, R. (2002). Element Concentrations in the Forest Moss *Hylocomium splendens*: Variation Associated with Altitude, Net Primary Production and Soil Chemistry. *Environmental Pollution*, Vol.116, No.1, (2002), pp. 129-135, ISSN 0269-7491

Gjengedal, E. & Steinnes, E. (1990). Uptake of Metal Ions in Moss from Artificial Precipitation. *Environmental Monitoring and Assessment*, Vol.14, No.1, (1990), pp. 77-87, ISSN 0167-6369

Harmens, H; Norris, D.A. Koerber, G.R. Buse, A. Steinnes, E. & Rühling, A. (2008a). Temporal Trends (1990 - 2000) in the Concentration of Cadmium, Lead and

Mercury in Mosses across Europe. *Environmental Pollution*, Vol.151, No.2, (2008), pp. 368-376, ISSN 0269-7491

Harrison, R.M. & Chirgawi, M.B. (1989). The Assessment of Air and Soil as Contributors of Some Trace Metals to Vegetable Plants. Use of a Filtered Air Growth Cabinet. *Science of the Total Environment*, **Vol.83**, No.1-2, (1989), pp. 13-34, ISSN 0048-9697

Kansanen, P. & Venetvaara, J. (1991). Comparison of Biological Collectors of Airborne Heavy Metals Near Ferrochrome and Steel Works. *Water, Air & Soil Pollution*, Vol.60, No.3-4, (1991), pp. 337-359, ISSN 0049-6979

Kovács, M. (1992a). Herbaceous (flowering) Plants. In: *Biological indicators in environmental protection*, M. Kovács, (Ed.), 76-82, Ellis Horwood Ltd, ISBN 0130849898, New York

Kovács, M. (1992b). Trees as Biological Indicators. In: *Biological indicators in environmental protection*. M. Kovács, (Ed.), 97-100, Ellis Horwood, ISBN 0130849898, New York.

Long, E.R.; MacDonald, D.D. Smith, S.L. & Calder, F.D. (1995). Incidence of Adverse Biological Effects within Ranges of Chemical Concentrations in Marine and Estuarine Sediments. *Environmental Management*, Vol.19, No.1, (1995), pp. 81-97, ISSN 0364-152X

MacNee, W. & Donaldson, K. (2000). How Can Ultrafine Particles be Responsible for Increased Mortality? *Monaldi Archives of Chest Disease*, Vol.55, No.2, (2000), pp. 135–139, ISSN 1122-0643

Mankovska, B.; Godzik, B. Badea, O. Shparyk, Y. & Moravcik, P. (2004). Chemical and Morphological Characteristics of Key Tree Species of the Carpathian Mountains. *Environmental Pollution*, Vol.130, No.1, (2004), pp. 41–54, ISSN 0269-7491

Manoli, E.; Voutsa, D. & Samara, C. (2002). Chemical Characterization and Source Identification Apportionment of Fine and Coarse Air Particles in Thessaloniki, Greece. *Atmospheric Environment*, Vol.36, No.6, (2002), pp. 949–961, ISSN 1352-2310

Markert, B. (1993). Instrumental Analysis of Plants, In: *Plants as biomonitors, indicators for heavy metals in terrestrial environment*, B. Markert, (Ed), 65–103, VCH, ISBN 1560812729, Weinheim

Markert, B. (2007). Definitions and Principles for Bioindication and Biomonitoring of Trace Metals in the Environment. *Journal of Trace Elements in Medicine and Biology*, Vol.21, S.1, (2007), pp. 77–82, ISSN 0946-672X

Markert, B.; Oehlmann, J. & Roth, M. (1997). General Aspects of Heavy Metal Monitoring by Plants and Animals, In: *Environmental biomonitoring - exposure, assessment and specimen banking*, K.S. Subramanian & G.V. Iyengar (Eds.), 19-29, ACS Symposium Series, Vol.654, American Chemical Society

Martin, M.H. & Coughtrey, P.J. (1982). *Biological Monitoring of Heavy Metal Pollution*, Kluwer Academic, ISBN 978-0853341369, London

Mielke, H.W.; Reagan, P.L. (1988). Soil as an Impact Pathway of Human Lead Exposure. *Environmental Health Perspectives*, Vol. 106, No. 1, (1988), pp. 217– 29, ISSN 0091-6765

Moore, C.C. (1974). A Modification of the Index of Atmospheric Purity Method for Substrate Differences. *The Lichenologists*, Vol.6, No.2, (1974), pp. 156-157, ISSN 0024-2829

Morselli, L.; Brusori, B. Passarini, F. Bernardi, E. Francaviglia, R. & Gatelata, L. (2004). Heavy Metal Monitoring at a Mediterranean Natural Ecosystem of Central Italy Trends in Different Environmental Matrixes. *Environment International*, Vol.30, No.2, (2004), pp. 173–181, ISSN 0160-4120

Mulgrew, A. & Williams, P. (2000). *Biomonitoring of Air Quality Using Plants*, WHO Collaborating Centre for Air Quality Management and Air Pollution Control. (WHO CC) ISSN 0938-9822, Berlin

Nali, C. & Lorenzini, G. (2007). Air Quality Survey Carried out by Schoolchildren: an Innovative Tool for Urban Planning. *Environmental Monitoring and Assessment*, Vol.131, No. 1-3, (2007), pp. 201-210, ISSN 0167-6369

Nash, T.H. (1988). Correlating Fumigation Studies with Field Effects, In: *Lichens, Bryophytes and Air Quality*, T.H. Nash, & V. Wirth, (Eds.), *Bibliotheca Lichenologica*, Vol.30, 201-216, J. Cramer, ISBN 3443580092, the University of Michigan

NG, O.H.; Tan, B.C. & Obbard, J.P. (2006). Lichens as Bioindicator of Atmospheric Heavy Metal Pollution in Singapore. *Environmental Monitoring and Assessment*, Vol.123, No.1-3, (2006), pp. 63-74, ISSN 0167-6369

Nriagu, J.O. (1980). Global Cycle and Properties of Nickel, In: *Nickel in the Environment*, J. O. Nriagu, (Ed.), 1-26, John Wiley & Sons, ISBN 0471058858, New York

Nurnberg, H.W.; Valenta, P. Nguyen, V.D. Godde, M. & Urano de Carralho, E. (1984). Studies on the Deposition of Acid and Ecotoxic Heavy Metals with Precipates from the Atmosphere. *Fresenius Journal of Analytical Chemistry*, Vol.317, No.3-4, (1984), pp. 314-323, ISSN 0937-0633

Økland, T.; Økland, R.H. & Steinnes, E. (1999). Element Concentrations in the Boreal Forest Moss *Hylocomium splendens*: Variation Related to Gradients in Vegetation and Local Environmental Factors. *Plant and Soil*, Vol.209, No.1, (1999), pp. 71-83, ISSN 0032-079X

Onianwa, P.C. (2000). Monitoring Atmospheric Metal Pollution: A Review of the Use of Mosses as Indicators. *Environmental Monitoring and Assessment*, Vol.71, No.1, (2000), pp. 13-50, ISSN 0167-6369

Pacheco, A.M.G. Barros, L.I.C. Freitas, M.C. Reis, M.A. Hipólito, C. & Oliveira, O.R. (2002). An Evaluation of Olive-tree Bark for the Biological Monitoring of Airborne Trace-Elements at Ground Level. *Environmental Pollution*, Vol.120, No.1, (2002), pp. 79-86, ISSN 0269-7491

Pakarinen, P. & Rinne, R.J.K. (1979). Growth Rates and Heavy Metals Concentrations of Five Moss Species in Paludified Spruce Forest. *Lindbergia*, Vol.5, No.2, (1979), pp. 77-83, ISSN 0105-0761

Panichev, N. & Mc Crindle, R.I. (2004). The Application of Bioindicators for the Assessment of Air Pollution. *Journal of Environmental Monitoring*, Vol.6, No.2, (2004), pp. 121-123, ISSN 1464-0325

Poikolainen, J. (2004). Mosses, Epiphytic Lichens and Tree Bark as Biomonitors for Air pollutants - Specifically for Heavy Metals in Regional Surveys. Faculty of Science, Department of Biology, University of Oulu; The Finnish Forest Research Institute, Muhos Research Station, ISBN 951-42-7479-2, Oulu, Finland

Prance, C.T.; Prance, A.E. & Sandved, K.B. (1993). *Bark: The formation, characteristics, and uses of bark around the world*, Timber Press, ISBN 978-0881922622, Portland

Puxbaum, H. (1991). Metal Compounds in the Atmosphere, In: *Metals and Their Compounds in the Environment: Occurrence, Analysis and Biological Relevance*, E. Merian, (Ed.), 257-286, Wiley-VCH, ISBN 352726521X, Weinheim

Rao, D.N. (1984). Response of Bryophytes to Air Pollution, In: *Bryophyte Ecology*, A.J.E. Smith (Ed), 445-471, Springer, ISBN 0412223406, London

Rasmussen, L. (1978). Element Content of Epiphytic *Hypnum cupressiforme* Related to Element Content of the Bark of Different Species of Phorophytes. *Lindbergia*, Vol.4, No.3-4, (1978), pp. 209-218, ISSN 0105-0761

Reimann, C.; Arnoldussen, A. Boyd, R. Finne, T. R. Nordgulen, Ø. Volden, T. & Englmaier, P. (2006). The Influence of a City on Element Contents of a Terrestrial Moss (Hylocomium splendens). *Science of the Total Environment*, Vol.369, No.1-3, (2006), pp. 419-32, ISSN 0048-9697

Richardson, D. H. S. (1981). *The Biology of Mosses*, Wiley, ISBN 9780470271902, USA

Richardson, D.H.S. & Nieboer, E. (1981). Lichens and Pollution Monitoring. *Endeavour*, Vol.5, No.3, (1981), pp. 127-133, ISSN 0160-9327

Rosman, K. J.; Ly, Ch. & Steinnes, E. (1998). Spatial and Temporal Variation in Isotopic Composition of Atmospheric Lead in Norwegian Moss. *Environmental Science & Technology*, Vol.32, No.17, (1998), pp. 2542-2546, ISSN 0013-936X

Ross, H.B. (1990). On the Use of Mosses (*Hylocomium splendens* and *Pleurozium schreberi*) for Estimating Atmospheric Trace Metal Deposition. *Water, Air, & Soil Pollution*, Vol.50, No.1-2, (1990), pp. 63-76, ISSN 0049-6979

Rossbach, M.; Jayasekera, R. Kniewald, G. & Hun, N. (1999). Large Scale Air Monitoring: Lichen vs. Air Particulate Matter Analysis. *Science of the Total Environment*, Vol.232, No.1-2, (1999), pp. 59-66, ISSN 0048-9697

Rossini Oliva, S. & Mingorance, M.D. (2006). Assessment of Airborne Heavy Metal Pollution by Aboveground Plant Parts, *Chemosphere*, Vol.65, No.2, (2006), pp. 177-182, ISSN 0045-6535

Seinfeld, J.H. & Pandis, S.N. (1998). *Atmospheric Chemistry and Physics: From Air Pollution to Climate Change*, Wiley-Interscience, ISBN 0471178160, New York, USA

Shakya, K.; Chettri, M. K. Sawidis, T. (2008). Impact of Heavy Metals (Copper, Zinc, and Lead) on the Chlorophyll Content of Some Mosses. *Archives of Environmental Contamination and Toxicology*, Vol.54, No.3, (2008), pp. 412-421, ISSN 0090-4341

Shrivastav, R. (2001). Atmospheric Heavy Metal Pollution (Development of Chronological Records and Geochemical Monitoring). Study in department of chemistry faculty of science, Dayalbagh Educational Institute, Agra, India. *Resonance*, 62-68. ISSN 0971-8044

Sloof, J. E. (1995). Lichens as Quantitative Biomonitors for Atmospheric Trace Element Deposition, Using Transplants. *Atmospheric Environment*, Vol.29, No.1, (1995), pp. ISSN 1352-2310

Smirnioudi, V.; Thomaidis, M.S. Piperaki, E.A. & Siskos, P.A. (1998). Determination of Trace Metals in Wet and Dust Deposition in Greece. *Fresenius Environmental Bulletin*, Vol.7, No.1, pp.85-90, ISSN 1018-4619

Smodiš, B. & Bleise, A. (2000). Biomonitoring of Atmospheric Pollution, *Proceeding of International Workshop on*, "*Internationally harmonised approach to biomonitoring trace element atmospheric deposition*", pp. 143–150, ISBN 92–0–100803–1, Portugal, 28 August–3 September, 2000

Spiegel, H. (2002). Trace Element Accumulation in Selected Bioindicators Exposed to Emissions along the Industrial Facilities of Danube Lowland. *Turkish Journal of Chemistry*, Vol.26, No.6, (2002), pp. 815 – 823, ISSN 1300-0527

Steinnes, E. (1993). Some Aspects of Biomonitoring of Air Pollutants Using Mosses, as Illustrated by the 1976 Norwegian Survey. In: *Plants as biomonitors. Indicators for heavy metals in the terrestrial environment*, B. Markert, (Ed), 381-394, VHC, ISBN 3527300015, Weinheim

Steinnes, E.; Johansen, O. Røyset, O. & Odegard, M. (1993). Comparison of Different Multi-Element Techniques for Analysis of Mosses Used as Biomonitors. *Environmental Monitoring and Assessment*, Vol.25, No.2, (1993), pp. 87-97, ISSN 0167-6369

Szczepaniak, K. & Biziuk, M. (2003). Aspects of the Biomonitoring Studies Using Mosses and Lichens as Indicators of Metal Pollution. *Environmental Research*, Vol.93, No.3, (2003), pp. 221–230, ISSN 0013-9351

Taylor, H.J.; Ashmore, M.R. & Bell, J.N.B. (1990). *Air Pollution Injury to Vegetation*, IEHO, ISBN 9780900103308, London

Turan, D.; Kocahakimoglu, C. Kavcar, P. Gaygısız, H. Atatanir, L. Turgut, C. & C.Sofuoglu, S. (2011). The Use of Olive Tree (Olea europaea L.) Leaves as a Bioindicator for Environmental Pollution in the Province of Aydın, Turkey. *Environmental Science and Pollution Research*, Vol.18, No.3, (2011), pp. 355–364, ISSN 0944-1344

W.H.O. (World Health Organization). (2002). Reducing Risks, Promoting Healthy Life. 1211 Geneva 27, Switzerland

Wannaz, W.D.; Carreras, H.B. Pérez, C.A. & Pignata, M.L. (2006). Assessment of Heavy Metal Accumulation in Two Species of Tillandsia in Relation to Atmospheric Emission Sources in Argentina. *Science of the Total Environment*, Vol.361, No.1-3, (2006), pp. 267–278, ISSN 0048-9697

Ward, N.I.; Reeves, R.D. & Brooks, R.R. (1975). Lead in Soils and Vegetation along a New Zealand State Highway with Low Traffic Volume. *Environmental Pollution*, Vol.9, No.4, (1975), pp. 243-251, ISSN 0269-7491

Watmough, S.A. & Hutchinson, T.C. (2003). Uptake of 207Pb and 111Cd through Bark of Mature Sugar Maple,White Ash and White Pine: a Field Experiment. *Environmental Pollution*, Vol.121, No.1, (2003), pp. 39–48, ISSN 0269-7491

Wittig, R. (1993). General Aspects of Biomonitoring Heavy Metals by Plants. In: *Plants as biomonitors - Indicators for heavy metals in the terrestrial environment*, B. Markert, (Ed), pp.3-27, VHC, ISBN 3527300015, Weinheim

Wolterbeek, B. (2002). Biomonitoring of Trace Element Air Pollution: Principles, Possibilities and Perspectives. *Environmental Pollution*, Vol.120, No.1, (2002), pp. 11–21, ISSN 0269-7491

Wolterbeek, H.Th. & Bode, P. (1999). Strategies in Sampling and Sample Handling in the Context of Large-scale Plant Biomonitoring Surveys of Trace Element Air Pollution. *Science of the Total Environment*, Vol.176, No.1-3, (1999), pp. 33-43, ISSN 0048-9697

Wolterbeek, H.Th. & Verburg, T.G. (2002). Judging Survey Quality: Local Variances. *Environtal Monitoring and Assessment*, Vol.73, No.1, pp. 7–16, ISSN 0167-6369

Wolterbeek, H.Th.; Garty, J. Reis, M.A. & Freitas, M.C. (2003). Biomonitors in Use: Lichens and Metal Air Pollution, In: *Bioindicators and biomonitors*, B.A. Markert, A.M. Breure, & H.G. Zechmeister, 377-419, Elsevier, ISBN 0080441777, Oxford

Zechmeister, H.G.; & Hohenwallner, D. (2006). A Comparison of Biomonitoring Methods for the Estimation of Atmospheric Pollutants in an Industrial Town in Austria. *Environmental Monitoring and Assessment*, Vol.117, No.1-3, (2006), pp. 245–259, ISSN 0167-6369

Zoller, W. H.; Gladney, E. S. & Duce R. A. (1974). Atmospheric Concentrations and Sources of Trace Metals at the South Pole. *Science*, Vol.183, No.4121, (1974), pp. 198-200, ISSN 0036-8075

7

On the Impact of Time-Resolved Boundary Conditions on the Simulation of Surface Ozone and PM10

Gabriele Curci

Dept. Physics, CETEMPS, University of L'Aquila
Italy

1. Introduction

The grid-spacing of chemistry-transport models (CTM) is always limited by computational resources and ranges from 100-200 km of global scale models, to 25-50 km of continental scale models, to 1-10 km of regional and local scale models. We push to higher resolution in hope of better reproducing small scale processes that affect our ability to assess the environmental and health impacts of emissions. Running simulations at a resolution less than 50 km is often feasible only using a limited area model, which uses a domain ascribed to the region of interest. However, even the air quality of a single city is in principle affected by all the emission sources at global level: we thus account for such long-range transport of pollutants and oxidants specifying the chemical state of the atmosphere outside the domain through boundary conditions (BC). BC concentrations are usually taken from typical profiles or from larger-scale simulations with a procedure called "nesting". In this chapter, we focus on the latter technique, exploring in particular the effect of different BC time-resolutions (monthly to hourly) on the simulation of ozone and particulate matter on a nested domain at European scale. For the sake of completeness, we point out here that even very high-resolution models cannot explicitly simulate processes at all possible spatial-temporal scales and thus a certain degree of parameterization is always required. For further insights on the "subgrid" issues we refer the reader to the literature (e.g. Galmarini et al., 2008; Qian et al., 2010; Denby et al., 2011; Paoli et al., 2011).

In addition to boundary conditions, chemistry-transport model simulations also require initial conditions (IC) for chemical species. The general aspects of the influences of IC and BC on the simulation may be understood in the simplified framework presented by Liu et al. (2001). The authors consider an Eulerian box model with one chemical species, whose evolution of concentration C is regulated by the species continuity equation:

$$\frac{dC}{dt} = P - LC + \frac{C_{BC} - C}{\tau_r} \qquad (1)$$

where P and L are the production and loss rates, respectively, C_{BC} is the background concentration, which represent the boundary condition in this case, and τ_r is the residence time into the box. The third term on the right-hand side isolates the source term attributable to the BC. The analytical solution of the equation is as follows (eq. 4 in Liu et al., 2001):

$$C(t) = C_{IC}\, e^{-(L+1/\tau_r)t} + \frac{P\tau_r}{L\tau_r + 1}\left(1 - e^{-(L+1/\tau_r)t}\right) + \frac{C_{BC}}{L\tau_r + 1}\left(1 - e^{-(L+1/\tau_r)t}\right) \qquad (2)$$

where C_{IC} is the initial condition. The influence of C_{IC} exponentially decreases with time, due both to photochemical loss and deposition (L) and outward-transport ($1/\tau_r$), thus it vanishes if a sufficient "spin-up" time is allowed. On the other hand, the importance of local sources (second term on the r.h.s.) and boundary conditions (third term on the r.h.s.) grow with time and drive the evolution of C after the "spin-up" time. The importance of BC is to be evaluated comparing the relative magnitudes of the local production term $P\tau_r$ (photochemical plus emission) against C_{BC}. If the local sources are much larger than C_{BC}, the influence of BC might be ignored. It is expected that boundary influence decreases during downwind transport and that it reaches a maximum when the arrival time is short and the species lifetime is long.

The simple considerations about the influence of IC and BC obtained from the solution in equation (2) of the box model, was demonstrated to be valid also for full three-dimensional Eulerian chemistry-transport models. Regarding IC, Berge et al. (2001) found that their influence a 3-D model decreases more slowly with time with respect to a box model, but still is reduced to <10% in the planetary boundary layer (PBL) after 3 days in a 400 km × 480 km domain covering Southern California. However, the same authors pointed out that the influence of IC might be >10% after 3 days for long lived grouped species (e.g. sum of reservoir species of ozone) and in the free troposphere. A similar spin-up time of 2 days for ozone in the PBL was reported by Jiménez et al. (2007) for a 272 km × 272 km covering North-Eastern Iberian Peninsula. On a larger domain covering Europe, Langmann & Bauer (2002) found that 5 days are needed by ozone in the PBL to "forget" its initial condition. More recently, a further sensitivity test at the North American continental scale, confirmed that a week is the minimum spin-up time recommended for a 3-D simulation of ozone and particulate matter (Samaali et al., 2009).

Liu et al. (2001) analyzed the influence of BC on their 3-D ozone simulation over California using the difference of concentrations between a reference run and another with zeroed boundary concentrations. They found that the percentage of ozone concentration attributable to BC is mostly determined by the distance from the domain edges, the influence being inversely proportional to the distance. The influence at a specific location and time is modulated by the characteristics of the local and upwind sources. During night the impact of BC on ozone is less than daytime, because of a less active photochemistry. It was calculated that BC may contribute 30-40% of ozone formation in polluted PBL, while stratospheric BC dominated ozone values in the free troposphere (Langmann & Bauer, 2002; Song et al., 2008). Barna & Knipping (2006) pointed out that a different representation of BC has a great impact on source-apportionment analysis.

While Liu et al. (2001) helped clarifying the general concepts of the influence of BC on chemistry-transport model simulations, several other studies focused on the impact of improved boundary conditions on simulations. Many studies found that increasing both temporal and spatial resolution of BC benefit the ozone (Langmann et al., 2003; Appel et al., 2007; Song et al., 2008; Szopa et al., 2009), carbon monoxide (Tang et al., 2007; Tombrou et al., 2009) and particulate matter simulation (Barna & Knipping, 2006; Borge et al., 2010). On the continental scale, BC have a significant impact on ozone background levels, while having much less impact on the variability and peak values (Tang et al., 2007; Szopa et al., 2009). In the free troposphere, a careful treatment of the variable tropopause is critical for a real advantage in using improved stratospheric ozone BC on model top (Lam and Fu, 2009; Makar et al., 2010).

The chapter is organized as follows. We first briefly describe the models used in the study in section 2. Then we study the impact and time scales of IC and BC on simulated ozone in section 3.1. In the same section, we analyse the difference among surface ozone simulations with the use of boundary conditions alternatively with monthly, daily or hourly update rate. In section 3.2 we analyse the effect of alternative BC on surface PM10, in particular during a Saharan dust event in July 2005. The scientific questions we shall try to address are:

- How long should be the spin-up time for simulated surface concentrations to be unaffected by initial conditions?
- How much is the contribution of boundary conditions to local ozone and PM10 levels?
- Is there any improvement in simulations of ozone and PM10 if the boundary conditions are provided at an higher frequency (up to the model time-step)?

In final section 4 we draw conclusions on these questions.

2. Models description

In this section we give a brief description of chemistry-transport models used in this study. We use the CHIMERE regional model (Bessagnet et al., 2008) to simulate lower atmosphere composition over continental Europe and the GEOS-Chem global model (Bey et al., 2001) to provide CHIMERE with gases and dust boundary conditions.

2.1 CHIMERE regional model

CHIMERE is a regional chemistry-transport model developed by a community of French institutions primarily designed to produce daily forecasts of ozone, aerosols and other pollutants and make long-term simulations for emissions control scenarios (Bessagnet et al., 2008; CHIMERE, 2011). In this study, The model is setup on a 0.5°×0.5° horizontal grid covering Europe (35°-58°N; 15°W-25°E) and 8 hybrid-sigma vertical layers extending to 500 hPa. Meteorological input is provided by PSU/NCAR MM5 model (Dudhia, 1993) run at 36×36 km² horizontal resolution and 29 vertical sigma layers extending up to 100 hPa, and regridded on the 0.5° × 0.5° CHIMERE grid. The model is forced by NCEP analyses using the grid nudging (grid FDDA) option implemented in MM5.

Anthropogenic emissions are derived from the Co-operative Programme for Monitoring and Evaluation of the Long-range Transmission of Air pollutants in Europe (EMEP) annual totals (Vestreng, 2003) for NOx, CO, VOC, SOx, NH3 and primary PM species, while carbonaceous aerosol emissions are taken from Junker & Liousse (2008). Biogenic emissions of isoprene and monoterpenes are calculated with the MEGAN model (Guenther et al., 2006). Dust and sea-salt emissions are also simulated on-line (CHIMERE, 2011).

The gas-phase chemical mechanism MELCHIOR (Latuatti, 1997) includes about 50 species and 120 reactions. The aerosol phase is simulated with a sectional approach with 8 size bins from 0.04 to 10 μm of diameter. The main processes governing the production and loss of main inorganic (sulphate, nitrate, ammonium) and organic secondary species are included.

Boundary conditions are implemented in a classical one-way approach. Species concentrations at the boundaries are introduced in the simulation through an outer envelope of model cells having the same resolution of the actual simulation grid. The boundary concentrations are transported inside the domain by the transport operator, i.e. the part of the model that simulates advection (Fig. 1). Regardless of the particular scheme choice (CHIMERE, 2011), the model uses information only from one upwind cells to solve for advection. Numerical stability is warranted by the adaptive time-step adjusted in order to have a Courant number always less than 1, e.g. for the zonal direction:

$$C_x = U \cdot \Delta t / \Delta x \qquad (3)$$

where C_x is the Courant number for the x-direction, U is the maximum zonal wind speed, Δt is the time step, and Δx is the grid spacing. By definition, the Courant number measures the influence that upwind concentrations may have on a given grid-cell in a single time-step in units of the grid size. If this number is less than one it means that information from only one upwind cell is needed. In CHIMERE, the time-step is adapted throughout the simulation to ensure this condition always holds. For some more details on the relationship among advection schemes and boundary conditions the reader is referred to the nice discussion given by Wang et al. (2004).

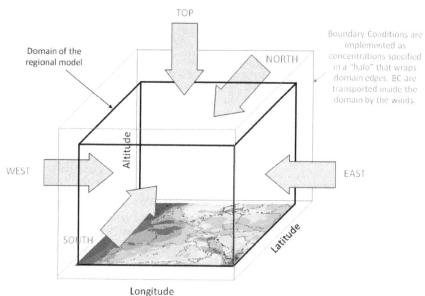

Fig. 1. Schematic of boundary conditions (BC) of a regional chemistry-transport model. The domain of simulation is denoted by the black cube, the 2-D map at the bottom is a sample of an output surface ozone field. Boundary concentrations of gases and aerosol species are passed to the model through an envelope of cells that wraps the domain around its edges. The transport operator of the model will use those cells for advection calculations and it will transport BC information into the domain.

In the default configuration, the boundary conditions for CHIMERE regional simulations are taken from monthly mean simulations of the global model LMDz-INCA (Hauglustaine et al., 2004) for species listed in Tab. 1. For aerosol species, size-resolved mass concentrations of global model are redistributed onto regional model size bins, using a simple linear interpolation in logarithmic bin diameters space. In Fig. 2 we show a sample of the monthly static ozone boundary conditions for the month of June.

CHIMERE species	Species long name	GEOS-Chem species
Gases		
O3	Ozone	O3
NO2	Nitrogen dioxide	NO2
HNO3	Nitric acid	HNO3
PAN	Peroxyacetylnitrate	PAN
H2O2	Hydrogen peroxide	H2O2
CO	Carbon monoxide	CO
CH4	Methane	-
HCHO	Formaldehyde	CH2O
C2H6	Ethane	C2H6
NC4H10	Butane and higher alkanes	ALK4
C2H4	Ethene	-
C3H6	Propene	PRPE
OXYL	Xylenes	-
Aerosol		
H2SO4	Sulfates	-
OC	Organic Carbon	-
BC	Black Carbon	-
DUST	Dust	DST1-4
SS	Sea Salts	-

Table 1. List of species for which boundary conditions are provided to CHIMERE regional chemistry-transport model. In the third column, the GEOS-Chem global model species used in this study to test sensitivity of CHIMERE to different boundary conditions.

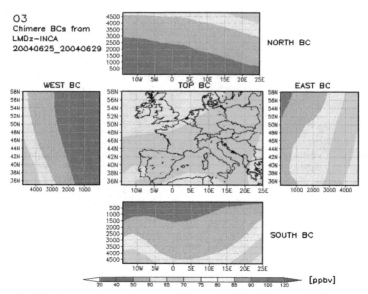

Fig. 2. Sample default ozone boundary conditions of CHIMERE model for the month of June. Ozone concentrations are taken from the monthly average of a simulation of the LMDz-INCA global model (Hauglustaine et al., 2004) and interpolated upon the CHIMERE horizontal and vertical grids. The panels show the resulting concentrations at the sides and top lid of the domain, as if the "box" of Fig. 1 has been open on a table.

2.2 GEOS-Chem global model

GEOS-Chem is a global chemistry model developed by a large international users community, originally stemming from the Harvard's Atmospheric Chemistry Modelling group (Bey et al., 2001; GEOS-Chem, 2011). The meteorological input is provided by the Goddard Earth Observing System (GEOS) of the NASA Global Modeling and Assimilation Office (GMAO). Mostly relevant to this study is the emission module developed by Fairlie et al. (2007), used here to include a more detailed contribution of Saharan dust emissions into the simulations. Desert dust emissions are lowered by a factor of three according to the study of Generoso et al. (2008).

GEOS-Chem simulations will be used in this study to produce alternative boundary conditions for the CHIMERE regional model at different time scales, from hourly to monthly.

3. Results

3.1 Ozone

3.1.1 Time-scales and impact of IC and BC

In order to study the effect of initial and boundary conditions on the CHIMERE ozone simulation at the European scale we use the simulations listed in Tab. 2. Basically, we alternatively zero IC and BC to isolate their effect through the difference with a reference simulation (Stein & Alpert, 1993). We choose a one month in summer (June 2005) in order to allow enough time to study the effect of IC and to ensure an active ozone photochemistry.

Simulation Label	Description
CTRL	Control simulation (reference)
NIC	No Initial Conditions (IC = 0)
NBC	No Boundary Conditions (BC = 0)

Table 2. List of simulations performed to study the effect of initial and boundary conditions on ozone.

The reference ozone simulation of CHIMERE is quickly evaluated against ground based measurements available from the EMEP network (www.emep.int). Let Obs_i^j and Mod_i^j be the observed and modeled values at time i and station j, respectively. Let N be the number of stations, and $Nobs^j$ the number of observations at station j.

- Pearson's Correlation (r) and coefficient of determination (R^2):

$$r = \frac{1}{N}\sum_{j=1}^{N}\frac{1}{Nobs^j - 1}\sum_{i=1}^{Nobs^j} Z_i^j(Mod)\cdot Z_i^j(Obs)$$

$$Z(X) = \frac{X - \langle X \rangle}{\sigma_X}$$

(4)

where X is a generic vector and $Z(X)$ is its standard score, also defined above. R^2 is defined as the square of r and denotes the fraction of variability of observations explained by the model.

- Mean Bias (MB):

$$MB = \frac{1}{N}\sum_{j=1}^{N}\left(\frac{1}{Nobs^j}\sum_{i=1}^{Nobs^j} Mod_i^j - Obs_i^j\right)$$

(5)

- Mean Normalized Bias Error (MNBE):

$$MNBE = \frac{1}{N}\sum_{j=1}^{N}\left(\frac{1}{Nobs^j}\sum_{i=1}^{Nobs^j}\frac{Mod_i^j - Obs_i^j}{Obs_i^j}\right)\times 100$$

(6)

- Mean Normalized Gross Error (MNGE):

$$MNGE = \frac{1}{N}\sum_{j=1}^{N}\left(\frac{1}{Nobs^j}\sum_{i=1}^{Nobs^j}\frac{|Mod_i^j - Obs_i^j|}{Obs_i^j}\right)\times 100$$

(7)

Results for current simulation are presented in Tab. 3 and Fig. 3. The model captures the central part of the observed distribution, but overestimates its low end and underestimates the upper end. This is a quite typical behaviour of regional chemistry-transport models (e.g. Appel et al., 2007). The average MNBE and MNGE for ozone hourly timeseries is slightly above the quality thresholds recommended by EPA (15% and 35% respectively), but the indices are well within the suggested limits for the daily maxima timeseries.

Variable	Observed mean	Modelled mean	MB	MNBE	MNGE	r
Units	$\mu g/m^3$	$\mu g/m^3$	$\mu g/m^3$	%	%	
O3 hourly	78.6	82.7	4.0	22.5	36.5	0.53
O3 daily max	103.4	98.7	-4.7	-0.4	15.8	0.69

Table 3. Comparison of observed and modelled ozone timeseries at EMEP monitoring stations. Values are averaged over all times and stations available for June 2005.

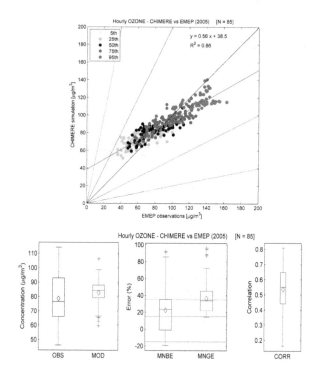

Fig. 3. Comparison of hourly ozone observed and modelled at EMEP stations. Left panel: scatter plot of 5th, 25th, 50th, 75th and 95th percentiles at all available stations. Right panel: boxplot of statistical indices shown in Tab. 3. Horizontal reference lines denote the quality level suggested by EPA (1991).

The impact of IC on ozone is studied through the difference between CTRL and NIC runs. We arbitrarily define the influence of Ψ_{IC} as the relative difference between CTRL and NIC simulations and the spin-up time τ_{IC} as the time needed to reduce Ψ_{IC} to less than 1%. In Fig. 4 we show the average ozone timeseries as measured and modelled at EMEP stations. After about 4 days of simulations the average Ψ_{IC} becomes negligible. According to our definition the average τ_{IC} is 3.9 days, but ranges from 0.5 to 8.1 days. In Fig. 5 we may appreciate the spatial distribution of τ_{IC}. We find a clear positive gradient from the North-West to the South-East of the domain. The reason may be found in the ozone distribution itself. In Fig. 6 we show the average ozone concentrations simulated by the model and we note that τ_{IC} gradients follows closely ozone gradients: the model just takes more time to build up ozone from the zero concentration starting point when the ozone level to be reached is higher. While this may pose questions on the method we used to estimate the "lifetime" of IC, this test is useful to verify that the model is essentially able to completely forget a whatever "wrong" initial condition after about 9 days of run.

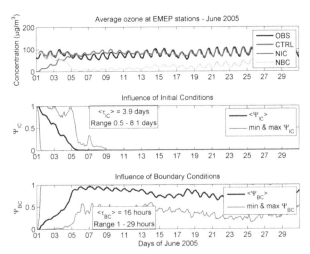

Fig. 4. Simulated influence of initial (IC) and boundary conditions (BC) in CHIMERE ozone at the European scale. The average timeseries at EMEP monitoring stations are shown. For explanation of simulation labels please refer to Tab. 2. The definition of Ψ_{IC}, Ψ_{BC}, τ_{IC}, and τ_{BC} is given in main text.

The influence of boundary conditions is studied in an analogous way. We define Ψ_{BC} as the difference between CTRL and NBC runs and the time of arrival of BC τ_{BC} as the first time when Ψ_{BC} is larger than 1%. In Fig. 4 we see that the behaviour of Ψ_{BC} mirrors that of Ψ_{IC}. As expected, the influence of BC grows as that of IC decreases. According to our definition, the average τ_{BC} is 16 hours, and ranges from 1 to 29 hours. As shown in Fig. 5, shortest times are found near the domain edges, while longest times are found in the interior of the domain. The variability inside the domain is attributable to the specific meteorological situations, since the time of arrival of BC at a specific location is determined by the winds encountered along the travel from the edges.

Until the spin-up time (τ_{IC}) is elapsed, Ψ_{BC} continues to ramp until a plateau is reached: the model is "warmed-up", and the ozone production at certain place is determined by the equilibrium between the sources inside the domain and the boundary conditions (eq. 2). For any time and location, Ψ_{BC} quantifies the relative influence of BC with respect to local production. The periods with higher values of Ψ_{BC} in timeseries of Fig. 4, thus indicate periods of less intense photochemical activity.

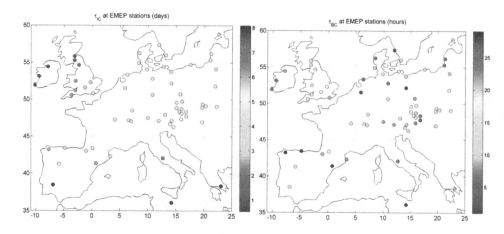

Fig. 5. Spatial distribution of the influence time of IC and BC on ozone.

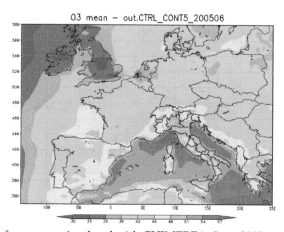

Fig. 6. Average surface ozone simulated with CHIMERE in June 2005.

The spatial distribution of the average Ψ_{BC} at the surface calculated from 10 to 30 June 2005 with CHIMERE is shown in Fig. 7. As also noted above, we find that the maximum influence of BC is around the borders, and it is striking to see that it is higher than 80-90% even in the polluted North-Western Europe. Ψ_{BC} reaches a minimum of less than 50% over Po Valley where the local production is invigorated by the elevated precursors emissions and very active photochemistry. This result imply that an error in BC may be effectively propagated into the simulation domain. For example, an error of 1 in the BC becomes 0.7 in a place where $\Psi_{BC} = 0.7$.

Liu et al. (2001) found that 3-D model results are consistent with analytical solution of a simple one-dimensional model, where the influence of BC may be written as:

$$\psi_{BC} = C_{BC} e^{-Lx/u} S_{x/u}(t) \tag{8}$$

where x is the distance to boundary and u is the wind speed, so that x/u is the time of BC arrival. S is a step function which is 0 for $t < x/u$ (before BC arrival) and 1 for $t \geq x/u$ (after BC arrival). If the ozone lifetime $\tau = 1/L$ is known, equation (8) can be applied directly to estimate Ψ_{BC}. Since the lifetime is generally unknown, but the BC arrival time τ_{BC} can be easily estimated, equation (8) can be inverted to roughly estimate the local averaged ozone lifetime. Inserting our definition of *relative* influence of BC on ozone Ψ_{BC} we obtain:

$$\tau = -\tau_{BC}/\log(\psi_{BC})$$ (9)

The result is shown in Fig. 7. Ozone lifetime estimated with this simple method ranges 1-3 days in the continental boundary layer and is longer than 5 days over the ocean, which is quite consistent with our expectations.

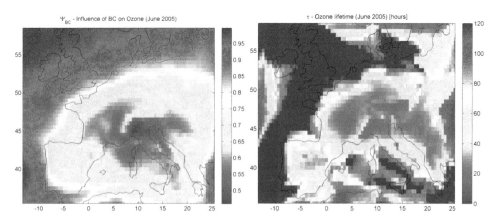

Fig. 7. Left: average influence of BC on ozone simulated in June 2005 (excluding first 9 days of spin-up). Right: ozone lifetime estimated with eq. 9.

3.1.2 Effect of alternative BC on surface ozone

We analyse the effect of different BC on simulated surface ozone using four simulations as listed in Tab. 4. We choose a longer summer period of two months, June and July 2005, to have a more robust statistics. Consistently with what found in previous section, we spin-up the model for 10 days.

Simulation Label	Description
CTRL	Control simulation (w/ reference LMDz-INCA BC)
BCGM	BC from GEOS-Chem monthly output
BCGD	BC from GEOS-Chem daily output
BCGH	BC from GEOS-Chem hourly output

Table 4. List of simulations performed to study the effect of alternative boundary conditions on ozone.

In Fig. 8 we compare the timeseries of ozone BC in the simulations averaged over the western border (leftmost rectangle in Fig. 2). The GEOS-Chem model simulates lower (higher) ozone values with respect to LMDz-INCA in June (July). One important reason, apart the many differences in models' formulation, is that the latter simulation is an average over five years of run, while GEOS-Chem simulates the "actual" (i.e. assimilated) meteorology of the CHIMERE simulation. The introduction of more detailed in time BC introduces much more variability, with differences up to ±30% with respect to the fixed monthly BC.

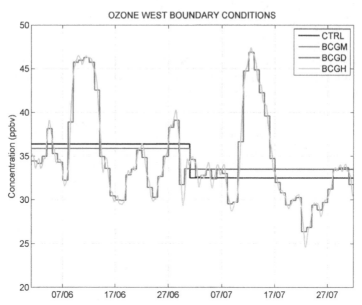

Fig. 8. Timeseries of the ozone boundary condition averaged over the west side of the European domain (left side of the box in Fig. 2). For explanation of the simulation labels please refer to Tab. 4.

In Tab. 5 we report the statistical summary, i.e. the average indices over all available EMEP stations, of the four simulations. The comparison with reference simulation over the two-month period is consistent with the one month simulation without spin-up (Tab. 3). The introduction of GEOS-Chem BC benefits the high CHIMERE model bias, reducing both the normalized bias and the gross error, probably because they are more specific of the simulation period than the LMDz-INCA climatology. The introduction of time resolution into the BC produces further reduction of model error and also significantly increases the correlation with the measurements.

Simulation	Observed mean	Modelled mean	MB	MNBE	MNGE	r
Units	$\mu g/m^3$	$\mu g/m^3$	$\mu g/m^3$	%	%	
CTRL	77.6	81.5	3.9	28.0	42.3	0.55
BCGM	77.6	77.2	-0.3	20.8	38.9	0.56
BCGD	77.6	75.4	-2.2	16.9	36.8	0.59
BCGH	77.6	75.3	-2.6	13.1	33.4	0.58

Table 5. Statistical indices of sensitivity simulations against EMEP hourly ozone measurements for June-July 2005. Values are averaged over all times and stations available.

In Fig. 9 we compare the simulated ozone with measurements at two selected EMEP stations, one near the border and very sensitive to BC (Mace Head, Ireland), the other about the centre of the domain and much less sensitive to BC (Heidenreichstein, Austria). For the Irish site we note interesting differences among the runs. During the first week of simulation, the monthly GEOS-Chem alternative BC enhances model underestimation with respect to reference, while the time-resolved GEOS-Chem BC slightly alleviates the bias with respect to reference. In the days around June 17th the time-resolved BC allow the model to capture a low ozone episode, but the subsequent week GEOS-Chem values are even too low. Also in other periods, the time-resolved BC allow the model to go closer to observations (July 7-10, 20-23). The correlation with measurements goes from 0.38 of the CTRL run, to 0.29 of the BCGM, to 0.44 of BCGD and BCGH. The gross error is reduced from 26% to 22% from CTRL to BCGD and BCGH runs. We also point out that the difference of the impact of hourly and daily BC is minimal.

The effect of alternative BC on the Austrian site, as expected is much less evident. However, the statistical indices of comparison with observation constantly get better as we introduce more resolution in time. The correlation increases from 0.56 to 0.61, the bias decreases from 23% to 14%, and the gross error decreases from 39% to 35%. Again, we note that using hourly or daily resolved BC does not significantly impact the simulation.

These results point out that the time resolution of BC may greatly affect the simulation, but an higher resolution may episodically worsen model skills. The big step is between monthly and daily resolved BC, while going down to hourly resolved BC, at the expense of more disk space and pre-processing time, does not yield further significant improvements..

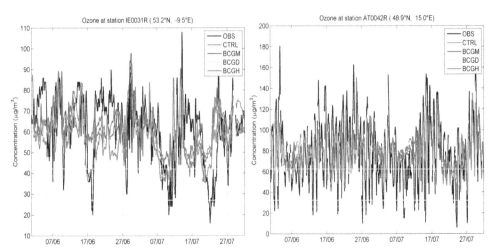

Fig. 9. Ozone timeseries in June-July 2005 as observed at two EMEP stations and simulated with CHIMERE with four sensitivity simulations (Tab. 4). Left: Mace Head station, close to the Western border of the domain and sensitive to BC (Ψ_{BC} ~0.95). Right: Heidenreichstein station, by the centre of the domain and less sensitive to BC (Ψ_{BC} ~0.63).

3.2 Aerosol

3.2.1 Effect of alternative BC on surface PM10

Similarly to ozone, we study the effect of alternative BC on simulated surface PM10 using the same simulations listed in Tab. 4. For this study, we choose to introduce alternative BC into CHIMERE only for dust, because the inflow of Saharan dust into the European domain is expected to contribute much more to the PM10 simulation than the other species (Curci et al., 2008; Curci & Beekmann, 2007). While transboundary pollution from anthropogenic sources is expected to impact background levels of PM (Park et al., 2003; 2004), Saharan dust may episodically yield to the exceedance of the PM10 limit for the protection of human health of 50 μg/m³ (Gobbi et al., 2007; Koçak et al., 2007; Perrino et al., 2009). It is estimated that in Italy the subtraction of natural dust to PM10 may yield to a reduction from 5% to 50% of the number of threshold exceedances depending on the meteorology and the station type (Pederzoli et al., 2010).

In Fig. 10 we compare the boundary conditions to CHIMERE from the Southern border in the four simulations. GEOS-Chem monthly mean dust values are higher than LMDz-INCA, and similarly to ozone the time-resolved BC have a much higher variability, with differences of ±50% and one episode with hourly values three times higher than monthly values.

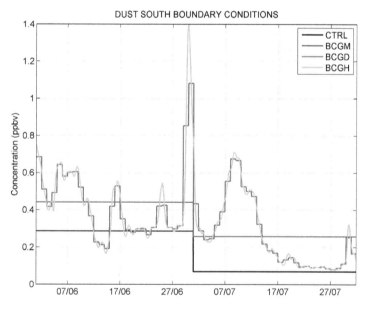

Fig. 10. Timeseries of the dust boundary condition summed over all CHIMERE size-bins and averaged over the south side of the European domain (bottom side of the box in Fig. 2). For explanation of the simulation labels please refer to Tab. 4.

In Tab. 6 we report the average statistical indices over all available EMEP stations of the four simulations. The higher dust values in GEOS-Chem BC drastically reduce the model bias from -40% to -7%, but the "noise" introduced by the large dust variability gradually degrades the correlation and the gross error as we increase the BC time resolution. This fact points out how tricky is the simulation of the Saharan dust contribution on European PM10 levels.

Simulation	Observed mean	Modelled mean	MB	MNBE	MNGE	r
Units	$\mu g/m^3$	$\mu g/m^3$	$\mu g/m^3$	%	%	
CTRL	17.5	9.0	-8.5	-41.9	47.4	0.63
BCGM	17.5	15.1	-2.4	-8.6	42.8	0.53
BCGD	17.5	16.0	-1.5	-6.9	50.2	0.50
BCGH	17.5	16.0	-1.5	-7.0	50.3	0.50

Table 6. Statistical indices of sensitivity simulations against EMEP daily PM10 measurements for June-July 2005. Values are averaged over all times and stations available.

In Fig. 11 we compare the PM10 timeseries at two EMEP stations, one near the South-Western border of the domain and more affected by Saharan dust (Barcarrota, Spain), and another to the North (Schauinsland, Germany). At the Spanish site, higher dust values in GEOS-Chem BC reduces the mean bias from -14 $\mu g/m^3$ to less than 1 $\mu g/m^3$, but the correlation decreases from 0.77 to 0.56 in CTRL and BCGH runs, respectively. The better resolved BC allow the model to better capture the observed PM10 variability by the end of the simulated period, but they also induce episodic overshoots during the first period of simulation that are completely unrealistic. Very similar features may be also noted at the German site, indicating that the importance of dust BC are not limited to the Southern part of the European domain.

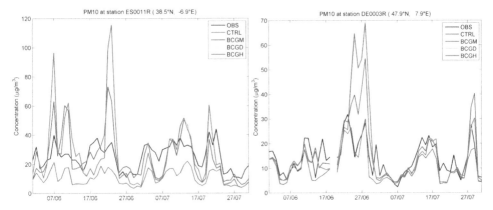

Fig. 11. PM10 timeseries in June-July 2005 as observed at two EMEP stations and simulated with CHIMERE with four sensitivity simulations (Tab. 4). Left: Barcarrota station, close to the South-Western border of the domain and strongly impacted by Saharan dust. Right: Schauinsland station, by the centre of the domain.

We now focus on the dust episode of 27-29 July 2005. In Fig. 12, the daily AOD observed by MODIS/Aqua between 24-29 July clearly tracks a dust cloud swapping Southern Europe from West to East. In Fig. 13 we show the same episode as recorded at ground level by the EMEP network and simulated by CHIMERE. The introduction of time-resolved BC helps the model in better reproducing both timing and magnitude of the event. Unfortunately, as we have seen in previous timeseries, this is not a general conclusions and further work is certainly warranted on the dust BC issue.

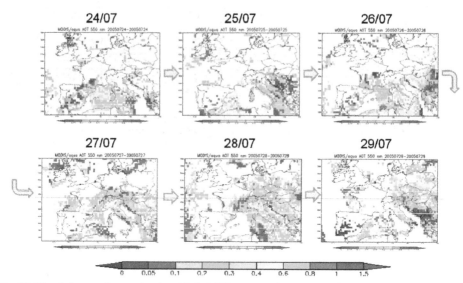

Fig. 12. The Saharan dust episode of July 2005 as seen from space through daily Aerosol Optical Depth (AOD) observations by MODIS/Terra.

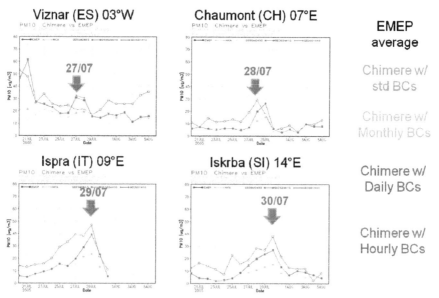

Fig. 13. The dust episode observed at EMEP stations and simulated with CHIMERE using different time-resolutions for the boundary conditions.

3.2.2 Eye-witness of a Saharan dust event over Central Italy

In a sort of "divine intervention", Sahara desert decided to produce one of its episodes while writing this chapter, during the first days of September 2011. The event was eye-witnessed by the author, and by its fellow citizens of L'Aquila in Central Italy, during the days 2-4. A large "bubble" of dust travelled over the South-Western Mediterranean and it was captured by the MODIS/Terra satellite instrument, as depicted in Fig. 14, which estimated a maximum optical thickness of more than 1.5. In Fig. 15 we show the striking effect on atmospheric visibility as observed from the ground: the mountain peak by the centre of the pictures, having a distance of about 25 km from the shot location, was almost obscured by the dust layer. The latter did actually hit the ground, as witnessed by the PM10 monitoring station in L'Aquila valley, which exceeded 50 µg/m^3 on September 3rd (Fig. 16), and also by the dust deposited over surfaces at the ground (e.g. the author's car depicted in Fig. 15).

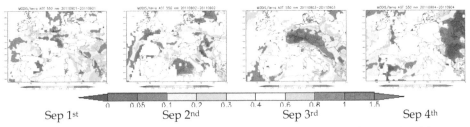

Sep 1st Sep 2nd Sep 3rd Sep 4th

Fig. 14. The Saharan dust event over Italy in September 2011 as seen from MODIS/Terra AOD observations.

Fig. 15. Left: view of L'Aquila valley in a clean summer day (10/09/2011); the mountain peak in the centre is at a distance of about 25 km. Middle: same view during a Saharan dust event (03/09/2011). Right: dust deposited over ground surfaces during the night.

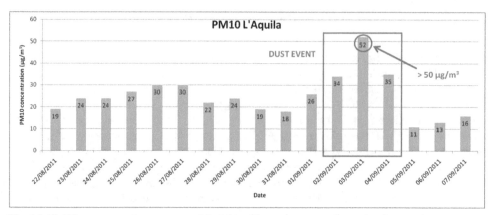

Fig. 16. PM10 concentration measured in L'Aquila at the ground monitoring station across the September 2011 Saharan dust event.

Interestingly, Fig. 17 shows that the arrival of the dust layer was qualitatively predicted by the ForeChem experimental chemical weather forecast system operating at University of L'Aquila (Curci, 2010), consisting of MM5/CHIMERE models automatically running, which is fed with the default monthly boundary conditions from LMDz-INCA global model (see Sec. 2.1).

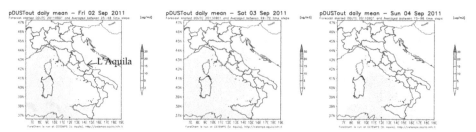

Fig. 17. Saharan dust event over Italy in September 2011 as forecasted with MM5/CHIMERE (ForeChem experimental chemical weather forecast operational at University of L'Aquila, http://pumpkin.aquila.infn.it/forechem/). Images show the fraction of PM10 at the ground due to dust from outside the domain, which is a nest within a European scale domain. Boundary conditions to the latter are provided with monthly mean fields from the LMDz-INCA global model.

4. Conclusion

The effect of initial (IC) and boundary conditions (BC) on the simulation of surface ozone and PM10 over the continental scale European domain is evaluated with several sensitivity tests of the CHIMERE chemistry-transport model.

Zeroing alternatively IC and BC in the model, and comparing the results with a reference run, we estimate an optimal model spin-up time of 9 days for the domain used in this study (35°-58°N; 15°W-25°E; 79 x 47 cells at 0.5° horizontal resolution). The BC have a significant impact on simulated ozone, especially in a belt of about 1000-2000 km around the domain borders. There, BC dominates ozone variability, while in the interior of the domain they have a weight similar to the local photochemical production. Through the BC test, the surface ozone lifetime is estimated to be 1-3 days over the continent and longer than 5 days over the oceans.

The impact of different time-resolution of BC is studied feeding the CHIMERE model with GEOS-Chem global model simulations. With respect to the reference BC, provided by five years monthly mean LMDz-INCA model simulations, the GEOS-Chem model has generally lower ozone and enhanced dust values during the period of simulation (June-July 2005). The positive ozone bias with respect to EMEP measurements is alleviated by GEOS-Chem BC, and also the introduction of BC better resoled in time benefit model skills. The average correlation increases from 0.55 to 0.59 and the normalized bias decreases from 28% to 17%. The large improvement is noticed when passing from monthly to daily BC, while hourly BC do not produce further improvements. We noticed, however, that time-resolved BC may episodically worsen model skills.

The introduction of different aerosol BC is tested focusing on dust, because of the prominent role that Saharan dust events play on the PM10 levels especially in Southern Europe. GEOS-Chem predicts higher dust concentrations with respect to LMDz-INCA, and its use as BC alleviates the CHIMERE low bias with respect to EMEP measurements. However, the agreement with observations get better during the events by the end of the simulated period (July 2005), but worsen at the beginning (June 2005). In particular, GEOS-Chem has a tendency to episodically overshooting dust at unrealistic levels. The introduction of time-resolved dust BC may allow the model to better reproducing both time and magnitude of Saharan events, but this is not a general conclusion with models' set-up used in this study. Further work on dust BC for European scale models is certainly needed in the future, possibly combining satellite observations with ground measurements of the aerosol composition and size-distribution that may better constrain the dust contribution to particulate matter.

5. Acknowledgment

The author was supported by the Italian Space Agency (ASI) in the frame of QUITSAT and PRIMES projects.

6. References

All figures in this chapter are originals by the author.

Appel, K. W.; Gilliland, A. B.; Sarwar, G. & Gilliam, R. C. (2007). Evaluation of the Community Multiscale Air Quality (CMAQ) model version 4.5: Sensitivities impacting model performance Part I—Ozone. *Atmospheric Environment*, Vol.41, pp. 9603–9615

Barna, M. G. & Knipping, E. M. (2006). Insights from the BRAVO study on nesting global models to specify boundary conditions in regional air quality modeling simulations. *Atmospheric Environment*, Vol.40, pp. S574–S582

Berge, E.; Huang, H.-C.; Chang, J. & Liu, T.-H. (2001). A study of the importance of initial conditions for photochemical oxidant modeling. *Journal of Geophysical Research*, Vol.106, pp. 1347-1363

Bessagnet, B.; Menut, L.; Curci, G.; Hodzic, A.; Guillaume, B.; Liousse, C.; Moukhtar, A.; Pun, B.; Seigneur, C. & Schulz, M. (2008). Regional modeling of carbonaceous aerosols over Europe - Focus on Secondary Organic Aerosols. *Journal of Atmospheric Chemistry*, Vol.61, pp. 175-202

Bey, I.; . Jacob, D. J.; Yantosca, R. M.; Logan, J. A.; Field, B.; Fiore, A. M.; Li, Q.; Liu, H.; Mickley, L. J. & Schultz, M. (2001). Global modeling of tropospheric chemistry with assimilated meteorology: Model description and evaluation. *Journal of Geophysical Research*, Vol.106, pp. 23073-23096

Borge, R.; López, J.; Lumbreras, J.; Narros, A. & Rodríguez, E. (2010). Influence of boundary conditions on CMAQ simulations over the Iberian Peninsula. *Atmospheric Environment*, Vol.44, pp. 2681-2695

CHIMERE (2011). The CHIMERE chemistry-transport model User's Guide, Available from: http://www.lmd.polytechnique.fr/chimere/chimere.php

Curci, G. (2010). An Air Quality Forecasting tool over Italy (ForeChem). In: *Proceedings of 31st Technical Meeting on Air Pollution Modelling and its Application*, Torino, Italy, Sep 2010

Curci, G. & Beekmann, M. (2007). Contribution of natural and biogenic emissions to the Air Quality in Europe. In: *NatAir final activity report*, pp. 141-175, Available from http://natair.ier.uni-stuttgart.de/NatAir_Final_Activity_Report.pdf

Curci, G.; Beekmann, M.; Vautard, R.; Bessagnet, B.; Menut, L.; Smiatek, G. ; Steinbrecher, G.; Theloke, J. & Friedrich, R. (2008). Impact of updated European biogenic emission inventory on air quality using Chimere chemistry-transport model. *IGAC 10th International Conference, Bridging the scales in Atmospheric Chemistry : Local to Global*, Annecy, France, September 7-12, 2008, Available from http://pumpkin.aquila.infn.it/gabri/downld/curci_igac2008_bioPM.pdf

Denby, B.; Cassiani, M.; de Smet, P.; de Leeuw, F. & Horálek, J. (2011). Sub-grid variability and its impact on European wide air quality exposure assessment. *Atmospheric Environment*, Vol.45, pp. 4220-4229

Dudhia, J. (1993). A nonhydrostatic version of the Penn State/NCAR mesoscale model: Validation tests and simulation of an Atlantic cyclone and cold front. *Monthly Weather Review*, Vol.121, pp. 1493–1513

EPA (1991). Guideline for regulatory application of the urban airshed model. U.S. Environmental Protection Agency, Office of Air Quality Planning and Standards, July 1991, 108 pp.

Fairlie, T. D.; Jacob, D. J. & Park, R. J. (2007). The impact of transpacific transport of mineral dust in the United States. *Atmospheric Environment*, Vol.41, pp. 1251-1266

Galmarini, S.; Vinuesa, J.-F. & Martilli, A. (2008). Modeling the impact of sub-grid scale emission variability on upper-air concentration. *Atmospheric Chemistry and Physics*, Vol.8, pp. 141-158

Generoso S.; Bey I.; Labonne M. & Breon F. M. (2008). Aerosol vertical distribution in dust outflow over the Atlantic: Comparisons between GEOS-Chem and Cloud-Aerosol Lidar and Infrared Pathfinder Satellite Observation (CALIPSO). *Journal of Geophysical Research*, Vol.113, D24209

GEOS-Chem (2011). The GEOS-Chem User's guide, Available from: www.geos-chem.org.

Gobbi, G. P.; Barnaba, F. & Ammannato, L. (2007). Estimating the impact of Saharan dust on the year 2001 PM10 record of Rome, Italy. *Atmospheric Environment*, Vol.41, No.2, (January 2007), pp. 261-275

Guenther, A.; Karl, T.; Harley, P.; Wiedinmyer, C.; Palmer, P. I. & Geron, C. (2006). Estimates of global terrestrial isoprene emissions using MEGAN (Model of Emissions of Gases and Aerosols from Nature). *Atmospheric Chemistry and Physics*, Vol.6, pp. 3181-3210

Hauglustaine, D. A.; Hourdin, F.; Walters, S.; Jourdain, L.; Filiberti, M.-A.; Larmarque, J.-F.; & Holland, E. A. (2004). Interactive chemistry in the Laboratoire de Météorologie Dynamique general circulation model: description and background tropospheric chemistry evaluation. *Journal of Geophysical Research*, Vol.109, D04314

Jiménez, P.; Parra, R. & Baldasano, J. M. (2007). Influence of initial and boundary conditions for ozone modelling in very complex terrains: A case study in the northeastern Iberian Peninsula. *Environmental Modelling & Software*, Vol.22, pp. 1294-1306

Junker, C. & Liousse, C. (2008). A global emission inventory of carbonaceous aerosol from historic records of fossil fuel and biofuel consumption for the period 1860–1997. *Atmospheric Chemistry and Physics*, Vol.8, pp. 1195–1207

Lam, Y. F. & Fu, J. S. (2009). A novel downscaling technique for the linkage of global and regional air quality modelling. *Atmospheric Chemistry and Physics*, Vol.9, pp. 9169–9185

Koçak, M.; Mihalopoulos, N. & Kubilay, N. (2007). Contributions of natural sources to high PM10 and PM2:5 events in the eastern Mediterranean. Atmospheric Environment Vol.41, 3806–3818

Langmann, B. & Bauer, S. E. (2002). On the Importance of Reliable Background Concentrations of Ozone for Regional Scale Photochemical Modelling. *Journal of Atmospheric Chemistry*, Vol.42, pp. 71–90

Langmann, B.; Bauer, S. E. & Bey, I. (2003). The influence of the global photochemical composition of the troposphere on European summer smog, Part I, Application of a global to mesoscale model chain. *Journal of Geophysical Research*, Vol.108, 4146 (14 pp.)

Lattuati, M. (1997). Contribution à l'étude du bilan de l'ozone troposphérique à l'interface de l'Europe et de l'Atlantique Nord: Modélisation lagrangienne et mesures en altitude, *Ph.D. thesis*, Univ. Pierre et Marie Curie, Paris, France

Liu, T.-H.; Jeng, F.-T.; Huang, H.-C. ; Berge, E. & Chang, J. S. (2001). Influences of initial conditions and boundary conditions on regional and urban scale Eulerian air quality transport model simulations. *Chemosphere – Global Change Science*, Vol.3, pp. 175–183

Makar, P. A.; Gong, W.; Mooney, C.; Zhang, J.; Davignon, D.; Samaali, M.; Moran, M. D.; He, H.; Tarasick, D. W.; Sills, D. & Chen, J. (2010). Dynamic adjustment of climatological ozone boundary conditions for air-quality forecasts. *Atmospheric Chemistry and Physics*, Vol.10, pp. 8997-9015

Paoli, R.; Cariolle, D. & Sausen, R. (2011). Review of effective emissions modeling and computation. *Geoscientific Model Development*, Vol.4, pp. 643-667

Park, R. J.; Jacob, D. J., Chin, M. & Martin, R. V. (2003). Sources of carbonaceous aerosols over the United States and implications for natural visibility. Journal of Geophysical Research, Vol.108, pp. 4355

Park, R. J.; Jacob, D. J., Field, B. D., Yantosca, R. M. & Chin, M. (2004). Natural and transboundary pollution influences on sulfate-nitrate-ammonium aerosols in the United States: Implications for policy, Journal of Geophysical Research, Vol.109, D15204

Pederzoli, A.; Mircea, M.; Finardi, S.; di Sarra, A. & Zanini G. (2010). Quantification of Saharan dust contribution to PM10 concentrations over Italy during 2003-2005. *Atmospheric Environment*, Vol.44, pp. 4181-4190

Perrino, C.; Canepari, S.; Catrambone, M.; Dalla Torre, S.; Rantica, E. & Sargolini, T. (2009). Influence of natural events on the concentration and composition of atmospheric particulate matter. *Atmospheric Environment*, Vol.43, pp. 4766-4779

Qian, Y.; Gustafson Jr., W. I. & Fast, J. D. (2010). An investigation of the sub-grid variability of trace gases and aerosols for global climate modelling. *Atmospheric Chemistry and Physics*, Vol.10, pp. 6917-6946

Samaali, M.; Moran, M. D. ; Bouchet, V. S. ; Pavlovic, R. ; Cousineau, S. & Sassi, M. (2009). On the influence of chemical initial and boundary conditions on annual regional air quality model simulations for North America. *Atmospheric Environment*, Vol.43, pp. 4873–4885

Song, C.-K.; Byun, D. W.; Pierce, R. B.; Alsaadi, J. A.; Schaack, T. K. & Vukovich, F. (2008), Downscale linkage of global model output for regional chemical transport modeling: Method and general performance. *Journal of Geophysical Research*, Vol.113, D08308 (15 pp.)

Stein, U. & Alpert, P. (1993). Factor separation in numerical simulations. *Journal of the Atmospheric Sciences*, Vol.50, pp. 2107-2115

Szopa, S.; Foret, G.; Menut, L & Cozic, A. (2009). Impact of large scale circulation on European summer surface ozone and consequences for modelling forecast. *Atmospheric Environment*, Vol.43, pp. 1189–1195

Tang, Y.; Carmichael, G. R.; Thongboonchoo, N.; Chai, T.; Horowitz, L. W.; Pierce, R. B.; Al-Saadi, J. A.; Pfister, G.; Vukovich, J. M.; Avery, M. A.; Sachse, G. W.; Ryerson, T. B.; Holloway, J. S.; Atlas, E. L.; Flocke, F. M.; Weber, R. J.; Huey, L. G.; Dibb, J. E.; Streets, D. G. & Brune, W. H. (2007). Influence of lateral and top boundary conditions on regional air quality prediction: A multiscale study coupling regional and global chemical transport models. *Journal of Geophysical Research*, Vol.112, D10S18 (21 pp.)

Tombrou, M.; Bossioli, E.; Protonotariou, A. P.; Flocas, H.; Giannakopoulos, C. & Dandou, A. (2009). Coupling GEOS-CHEM with A Regional Air Pollution Model for Greece. *Atmospheric Environment*, Vol.43, pp. 4793–4804

Vestreng, V. (2003). Review and revision. Emission data reported to CLRTAP, Tech. rep., EMEP MSC-W

Wang, Y. X.; McElroy, M. B.; Jacob, D. J. & Yantosca, R. M. (2004). A nested grid formulation for chemical transport over Asia: Applications to CO. *Journal of Geophysical Research*, Vol.109, D22307 (20 pp.)

Winner, D. A. & Cass, G. R. (1995). Effect of alternative boundary conditions on predicted ozone control strategy performance: A case study in the Los Angeles area. *Atmospheric Environment*, Vol.29, pp. 3451-3464

8

Recovery and Reuse of SO_2 from Thermal Power Plant Emission

Arun Kumar Sharma, Shveta Acharya,
Rashmi Sharma and Meenakshi Saxena
Department of Chemistry,
S.D. Govt. College, Beawar, Rajasthan
India

1. Introduction

Thermal power plants are major sources of air pollutants. Three major air pollutants emitted from thermal power plant are Suspended Particulate Matter (SPM), Sulphur di oxide – (SO_2), Sulphur tri oxide (SO3), and (NO2, NO3). The amount of pollutants emitted from any power plant depends upon the type of the fuel used, burning method and type of control equipment. These pollutants finally found in ambient air. Coal is re-emerging as a the dominant fuel for power generation in various power plants. [1] Various coal such as petcock, lignite, bituminous etc. used in power plants in which % S have 6.0 %,4.0 % and 3.8 % respectively. The common elements in fuel are Carbon, which is principle combustible constitute of all fossils. Oxygen, nitrogen, hydrogen, and Sulphur (S) are not combustible elements. Sulphur in coal cannot be destroyed it can only be converted to one form to another During the combustion process , Sulphur react with oxygen and formed SO_2 and SO_3.

SO_2 is a major constituent in air pollution.[2] and affects the environment by no. of ways like acid rain, corrosions and severe damage to the health. SO_2 causes a wide variety of health and environmental impacts because of the way it reacts with other substances in the air. Particularly sensitive groups include people with asthma who are active outdoors and children, the elderly, and people with heart or lung disease. Intensity of SO_2 emission can be observed by following example. " A typical 6 MW power generation unit using furnace oil containing 2 % Sulphur will emit 388 tons of SO_2 per year, based upon 320 working days or A 22.5 MW power generation unit will emit 1690 tons of SO_2 per year by using Pet Coke."[3-4]

2. Review of literature

The acid rain problem is mainly attributed to anthropogenic sulphur dioxide and, to a lesser extent, nitrogen oxide emissions. Sulphur dioxide can be directly removed from the atmosphere through dry or wet deposition. The main sink of atmospheric SO_2, however, is the oxidation to SO_4^{2-} in the gas phase and in the liquid phase of clouds, fog and rain. These processes are, besides the oxidation of nitrogen compounds, the major source of acidity in acid rain or acid fog. The relative importance of different pathways for atmospheric oxidation of sulphur dioxide can vary under different conditions such as relative humidity, intensity of solar radiations, temperature and degree of air pollution. Under favourable conditions the oxidation of sulphur dioxide can occur in the atmospheric aqueous phase at significantly faster rates than in the gas phase. It is believed that, on a global scale, more than 70% of the global oxidation of SO_2 to SO_4^{2-} occurs within cloud droplets (Langner and Rodhe 1991).The oxidation of sulphur dioxide has been one of the most frequently studied reactions in aqueous atmospheric droplets. Three reaction pathways are considered to be dominantly responsible for oxidation of SO_2 in atmospheric water droplets. These are the oxidation of dissolved SO_2 by H_2O_2, O_3 and O_2 in the presence of transition metal ions as catalysts (Seinfeld and Pandis 1998; Warneck et al. 1996). In acid solutions the major oxidant is H_2O_2, whereas the role of O_3 becomes more important above pH 6 (Calvert et al. 1985; Ibusuki et al. 1990; Seinfeld and Pandis 1998). Oxidation by molecular oxygen may also be important if cloud water contains sufficient amount of transition metal ions (e.g., Fe, Mn) for autocatalytic reactions to occur. This process may play an important role in highly industrialized areas where various transition metals are present in atmospheric water in relatively high concentrations (Seigneur and Saxena 1984). Catalytic autoxidation of S(IV) is the subject of a number of studies (e.g. Penkett et al.1979; Pasiuk-Bronikowska and Bronikowski 1981; Martin 1984; Martin and Hill 1987; Ibusuki and Takeuchi 1987; Kraft and van Eldik 1989; Grgić et al. 1991, 1992; Berglund and Elding 1995; Novič et al. 1996; Turšič et al. 2003). It is claimed that at pH 4, transition metal catalysed pathways could account for up to half of the oxidation of S(IV) to S(VI) (Graedel et al. 1985). According to present knowledge, iron(II/III) and manganese (II/III) are the most important catalysts in atmospheric droplets (Coichev and van Eldik 1994;Brandt and van Eldik 1995; Seinfeld and Pandis 1998). These metals are the only efficient catalysts at low pH. In addition, both iron and manganese are common constituents of tropospheric aerosols and water droplets even in remote areas due to their generation from erosion of the earth's crust. Other transition metals such as Cu(II), Co(III), Sc(III), Ti(III), V(III) and Cr(III), are also catalysts, but with a substantially lower effect on the reaction rate (Ibusuki et al. 1990; Grgić et al. 1991; Sedlak and Hoigné 1993). The catalytic oxidation of S(IV) is a free radical chain reaction. Its mechanism and kinetics are so complex and sensitive to the conditions under which the process occurs that even a minor change in experimental conditions can cause a change of the dominant path of the reaction course, and thus lead to diverse results. Despite numerous studies of the metal catalysed S(IV) oxidation there still exist serious discrepancies in rate expressions, rate constants, pH dependencies, activation energies, reaction mechanisms etc. Recent studies show that the sulphur(IV) oxidation in atmospheric water droplets can be affected by other reactions. In particular, organic chemistry may be especially important. Organic compounds may dissolve into water droplets and react with sulphoxy radicals and transition metal ions, and thus alter the rate of catalytic S(IV) oxidation (Martin et al. 1991; Pasiuk-Bronikowska et al. 1997; Grgić et al. 1998; Pasiuk-Bronikowska et al. 2003a,b; Ziajka and Pasiuk-Bronikowska 2003, 2005).Recently, the inhibiting effect of such organic ligands as oxalate, acetate and formate in the iron-catalyzed autoxidation of sulphur(IV) oxides in atmospheric water droplets has been suggested. Grgić et al. (1998, 1999) and Wolf et al. (2000) reported the strong inhibiting effect of oxalate on the Fe-catalysed S(IV) oxidation in aqueous acidic solution. Acetate and formate also inhibit the reaction, but to a much lesser extent than oxalate (Grgićet al. 1998). Very recently, the influence of some low weight mono- (formic, acetic, glycolic,lactic) and di-carboxylic acids (oxalic, malic, malonic) on the Mn(II)-

catalysed S(IV) oxidation has also been investigated (Grgić et al. 2002; Podkrajšek et al. 2006). It has been established that mono-carboxylic acids inhibit the oxidation, with the strongest influence 2 J Atmos Chem (2008) 60:1–17 found for formic acid. The lowest inhibition was caused by acetic acid. From among dicarboxylic acids, oxalic acid slows down the S(IV) oxidation, although to a lesser extent than mono-carboxylic acids, while malic and malonic acids have practically no influence. The effect of organic compounds in atmospheric water on the transition metal-catalysed oxidation of sulphur(IV) is not fully known yet and more work in this area is needed to understand these processes better. The purpose of the present study was to study the kinetics of the Mn(II)-catalysed S(IV) oxidation and to determine the inhibiting effect of acetic acid on this process under different experimental conditions representative for heavily polluted areas. The experiments were carried out at Mn(II) and CH_3COOH concentrations in the range 10^{-6}–10^{-5} and 10^{-6}–10^{-4} mol/dm^3, respectively, and at initial pH of the solution in the range 3.5–5.0; initial concentrations of S(IV) were around 10^{-3} mol/dm^3. S(IV) liquid-phase concentration of 1×10^{-3} mol/dm^3 corresponds to 0.6 ppm SO_2 in the gas phase over a solution of pH=5, or 7 ppm over a solution of pH=4, or 20 ppm over a solution of pH=3.5. Such high SO_2 gas-phase concentrations are found in heavily polluted areas as well as in power plant and volcanic plumes. In highly polluted locations, for example in large urban areas where coal is used for domestic heating purposes, or for poorly controlled combustion in industrial installations, SO_2 concentrations are rather high and vary between 0.1 and 0.5 ppm, and sometimes they are even higher (Ferrari and Salisbury 1999). High sulphur dioxide concentrations are being recorded in some of the megacities in developing countries where burning of coal is the main source of energy. The greatest problems related to sulfur dioxide occur in Asia (mainly in Chinese cities and some Middle-East cities such as Teheran, Tbilisi and Istanbul)(Baldasano et al. 2003). In Asia there are cities [e.g. Guiyang (424 µg/m3), Chongquing (340 µg/m3)] with average annual values of more than six times the WHO guideline value (Baldasano et al. 2003). Also in Africa some of the urban areas, and especially industrial areas, experience high concentrations of sulphur dioxide (WHO 2006). Weekly average concentrations in Zambia's copper belt (Nkana, Mufulira and Luanshya) were found to range from 167 to 672 µg/m3, the highest weekly average being 1,400 µg/m3. Studies undertaken on the impact of the Selebi Phikwe copper smelter in Botswana show that there are large areas experiencing concentrations above 100 µg/m3. Short term measurement indicated 1-h average concentrations of more than 1,000 µg/m3 (WHO 2006). Also some of the heavily industrialized areas in Europe may still be experiencing high levels of sulphur dioxide. In some cities in the north western corner of the Russian Federation, close to large primary smelters, daily concentrations of sulphur dioxide exceed 1,000 µg/m3 (WHO 2006). From the point of view of atmospheric chemistry, especially fast chemical reactions, concentrations averaged for shorter periods e.g., for 1 h or even for several minutes, are more relevant. These concentrations are closer to actual concentrations at which fast reactions proceed in the atmosphere. Concentrations averaged for shorter periods are considerably higher than those averaged for longer periods. Peak concentrations over shorter averaging periods may still be very high, both in cities with a high use of coal for domestic heating and when plumes of effluent from power station chimneys fall to the ground (fumigation episodes). Transient peak concentrations of several thousand

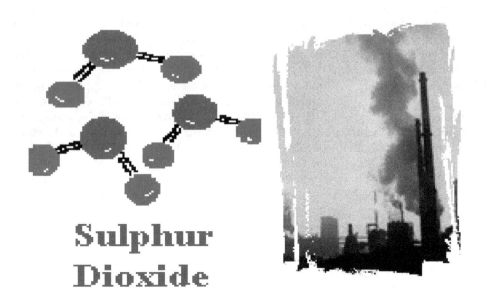

Snap. 1. Photographs showing air pollution by industries along with molecular structure of SO_2.

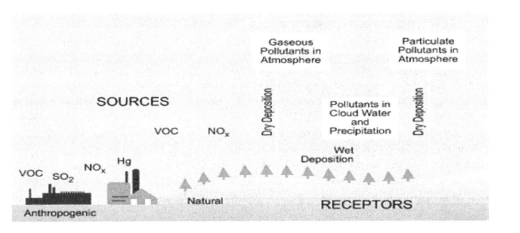

Snap. 2. Photographs showing Acid rain effect of SO$_2$.

Snap. 3. Photographs showing Acid rain effect of SO$_2$ on monuments of India.

microgram per cubic meter are not uncommon (WHO 2000). Concentrations of SO_2 can reach tens of parts per million in power plant (Jaakkola et al.1998) and volcanic plumes (Gauthier and Le Cloarec 1998; Shinohara 2005). Under stable atmospheric conditions, SO_2 may be transported relatively great distances at appreciable concentrations. In very stable power station plumes SO_2 concentrations may be greater than 1.0 ppm over 70 km from their source (Stephens and McCaldin 1971). J Atmos Chem (2008) 60:1-17 3 Concentrations of Mn(II) and acetic acid in solutions used in our experiments correspond to those found in rain-, cloud- and fogwater in heavily polluted urban and industrialized areas. Manganese is one of the most abundant transition metals in atmospheric liquid phases (wet aerosol, cloud, fog, rain). The only source of these metals in the atmospheric aqueous phase is the dissolution of aerosol particles incorporated in water droplets. The common particles containing trace metals are emitted from both anthropogenic (fossil fuel combustion, industrial processes) and natural (windblown dust, weathering, volcanoes) sources. Particles from anthropogenic sources contribute significantly to metal distribution in atmospheric droplets due to their high metal content and solubility. In consequence, trace metal concentrations in atmospheric waters are higher in urban and industrial areas (Colin et al. 1990). In atmospheric waters, manganese is mainly found as Mn(II), which is more soluble than manganese(III) (Deutsch et al. 1997). Mn(II) exhibit a large variation in solubility dependent on the nature of the particles, but this solubility is pH-independent (Millet et al.1995). Concentrations of Mn in rainwater are typically lower than those observed in fog and cloud water samples. In urban and industrial areas manganese concentrations range from 10^{-7} to 10^{-6} mol/dm3 in rain (Deutsch et al. 1997; Patel et al. 2001), and from 10^{-6} to 10^{-5} mol/dm3 in fog- and cloudwater (Millet et al. 1995; Brandt and van Eldik 1995) Acetic acid is one of the most abundant carboxylic acids in the troposphere and it is found in rain, clouds, fogs, and aerosol particles from remote to highly polluted urban areas. The atmospheric sources of carboxylic acids are numerous and they comprise: primary biogenic emissions, primary anthropogenic emissions and photochemical transformations of precursors in aqueous, gaseous, and particulate phases (Chebbi and Carlier 1996). Direct anthropogenic emissions of carboxylic acids (e.g. from incomplete combustion of fossil fuels, wood and other biomass material) and/or photo-oxidation of anthropogenic organic compounds are the main sources of these compounds in urban and industrial environments (Chebbi and Carlier 1996; Kawamura et al. 1996). Concentrations of organic acids are generally elevated in the urban as compared with the nonurban atmosphere (Meng and Seinfeld 1995). In urban sites concentrations of acetic acid range from 10^{-7} to 10^{-5} mol/dm3 in rain (Chebbi and Carlier 1996; Kawamura et al. 1996), and from 10^{-6} to 10^{-4} mol/dm3 in cloud and fogwater (Meng and Seinfeld 1995; Brandt and van Eldik 1995; Millet et al. 1996; Raja et al.

3. Basics and scopes of the work – Flue Gas Desulphurization (FGD)

Flue gas desulphurization (FGD) is the current state-of-the art technology used for removing sulphur dioxide from the exhaust flue gases in power plants. SO_2 is an acid gas and thus the typical sorbent slurries or other materials used to remove the SO_2 from the flue gases are alkaline. The reaction taking place in wet scrubbing using Ca (OH) $_2$ and NaOH slurry produces $CaSO_3$ and Na_2SO_3 and can be expressed as:

$$Ca(OH)_2 \text{ (solid)} + SO_2 \text{ (gas)} \longrightarrow CaSO_3 \text{ (solid)} + H_2O \text{ (liquid)}$$

$$2NaOH \text{ (solid)} + SO_2 \text{ (gas)} \longrightarrow Na_2SO_3 \text{ (solid)} + H_2O \text{ (Liquid)}$$

Some FGD systems go a step further and oxidize the $CaSO_3$ and Na_2SO_3 to produce marketable $CaSO_4 \cdot 2H_2O$ (gypsum) and Na_2SO_4 (Sodium Sulphate): [5-6]

$$CaSO_3 \text{ (solid)} + \tfrac{1}{2}O_2 \text{ (gas)} + 2H_2O \text{ (liquid)} \longrightarrow CaSO_4 \cdot 2H_2O \text{ (solid)}$$

$$Na_2SO_3 \text{ (solid)} + \tfrac{1}{2} O_2 \text{ (gas)} \longrightarrow Na_2SO_4 \text{ (solid)}$$

3.1 Mechanism

When sulfur dioxide (SO_2) in the flue gas contacts scrubber slurries, the pollutant transfers from the gas to the liquid phase, where the following equilibrium reactions are fundamentally representative of the transfer process.

$$SO_2 + H_2O \longrightarrow H_2SO_3 \longrightarrow H^+ + HSO3^- \longrightarrow H^+ + SO_3^{-2} \qquad (1)$$

when lime hydrated powder or caustic flakes introduced to water will raise the pH according to the following mechanism.

$$Ca(OH)_2 \longrightarrow Ca^{+2} + 2\,OH^{-1} \qquad (2)$$

$$NaOH \longrightarrow Na^{+1} + OH^{-1}.. \qquad (2.1)$$

However, $Ca(OH)_2$ is only slightly soluble in water, so this reaction is minor in and of itself. In the presence of acid, calcium hydroxide reacts much more vigorously and it is the acid generated by absorption of SO_2 into the liquid that drives the lime dissolution process.

$$Ca(OH)_2 + 2H^+ \longrightarrow Ca^{+2} + 2\,H_2O \qquad (3)$$

$$NaOH + H^+ \longrightarrow Na^+ + H_2O \qquad (3.1)$$

Equations 1, 2 and 3 when combined illustrate the primary scrubbing mechanism.

$$Ca(OH)_2 + 2\,H^+ + SO_3^{-2} \longrightarrow Ca^{+2} + SO_3^{-2} + 2H_2O \qquad (4)$$

$$2NaOH + H^+ + SO_3^{-2} \longrightarrow Na_2SO_3 + H_2O + OH^- \qquad (4.1)$$

In the absence of any other factors, (for example, oxygen in flue gas) calcium and sulfite ions will precipitate as a hemihydrate, where water is actually included in the crystal lattice of the scrubber byproduct.

$$Ca^{+2} + SO_3^{-2} + \tfrac{1}{2} H_2O \longrightarrow CaSO_3.1/2\ H_2O.. \qquad (5)$$

However, oxygen in the flue gas has a major effect on chemistry, and in particular on byproduct formation. Aqueous bisulfite and sulfite ions react with oxygen to produce sulfate ions (SO_4-2).

$$2 SO_3^{-2} + O_2 \quad\xrightarrow{\hspace{3cm}}\quad 2 SO_4^{2-}... \tag{6}$$

Approximately the first 15 mole percent of the sulfate ions co-precipitate with sulfite to form calcium sulfite-sulfate hemihydrate $[(CaSO_3 \cdot CaSO_4) \cdot \frac{1}{2}H_2O]$. Any sulfate above the 15 percent mole ratio precipitates with calcium as gypsum.

$$Ca^{+2} + SO_4^{2-} + 2H_2O \quad\xrightarrow{\hspace{3cm}}\quad CaSO_4.2H_2O \tag{7}$$

$$2Na^+ + SO_4^{2-} \quad\xrightarrow{\hspace{3cm}}\quad Na_2SO_4... \tag{8}$$

Calcium sulfite-sulfate hemihydrate is a soft, difficult-to-dewater material that previously has had little practical value as a chemical commodity. Gypsum, on the other hand, is much easier to handle and has practical value. These factors are driving utilities to install forced oxidation systems for gypsum production.

There are three control technologies which have major application in the field of Sulphur di Oxide control.[7-8]

- Adsorption.
- Catalytic Oxidation / reduction.
- Absorption.

Adsorption is a control technology for control of SO_2 from stack gases but suffers from several following drawbacks viz:

1. Higher energy requirements.
2. Penetration of SO_2 in the granule is difficult.
3. Highly active absorbent surfaces cause oxidation of SO_2 to SO_3 which react with moisture in flue gases to form acid.
4. Regeneration techniques are costlier.

Catalytically oxidation / reduction is a control technology for control of from stack gases but suffers from several following drawbacks viz:

1. Higher energy requirements
2. Large equipment size.
3. Costly Catalysts.
4. Regeneration and disposal of catalysts is also a problem
5. Contractor design is complex.

Absorption is a control technology for control of SO_2 from stack gases is most widely practiced.

However this technology also suffers from following drawbacks:

1. Stack gas cooling and reheating is required.
2. Mist elimination is required.

However these problems can be easily encountered with proper engineering design used. Besides this less operator's intensiveness, less cost and ease of handling of liquid sorbent makes it an attractive option. It is one of the most widely used control technology employed for removal of SO_2 [9-10]

4. Material and methods

All experiments were conducted on Stack monitoring Kit (Model No. and Make -VSS1, 141 DTH -2005,Vayubodhan). First of all Stack monitoring kit of SO_2 monitoring were set up for experiment at *chimney* inlet of Boiler of thermal power plant. Flue gas containing SO_2 were supplied from *chimney* via probe connected with flexible pipe of stack monitoring kit. The flow of flue gas were controlled using an inlet line Rota meter and was maintained at a value of 3 liter per minute and other end of flexible pipe carrying air and SO_2 respectively were connected to a impinger of 10 cm diameter and 100 cm length. The impinger were filled with 100 ml of scrubbing media in this experiment i.e. Sludge solution, Calcium hydroxide solution, Sodium hydroxide solution.

The concentration of SO_2 in flue gases was first measured by Stack monitoring Kit.

$$C_{SO2} = \frac{K_2 \, (V_t - V_{tb}) \, N \, (V_{soln})}{V_{m(std)} * V_a}$$

C_{SO2} = Concentration of sulphur dioxide, dry basis converted to standard conditions, mg/NM^3.

N = Normality of barium per chlorate titrant mili equivalent/ml.

K_2 = 32.03 mg/meq.

V_t = Volume of barium per chlorate titrant used for the sample, ml.

V_{tb} = Volume of barium per chlorate titrant used for the blank, ml

$V_{m(std)}$ = Dry gas volume measured by the dry gas meter, corrected to standard conditions, NM^3.

V_a = Volume of sample aliquot-titrated, ml.

Five sets of reading were taken by varying concentration of every solution. 100 ml of solution were taken in first two different impinges for better absorption of SO_2 and 30 ml of H_2O_2 was taken in the third for determination of remaining SO_2. Respective sulphate were formed in solution. Dissolved sulphate were extracted from solution by heating till dryness. Three parameters regards to % SO_3 (gravimetric), % SO_2 (Volumetric) and % alkalinity were analyzed in precipitate. The methods used as Indian standard method from bureau of Indian standard.[11-16] During the experiments pipette out 10 ml of NaOH solution in every 15 minutes and   pH were analyzed, titrate with 1M oxalic acid determination for fall in conc. of NaOH. Similarly Experiments were conducted on Indirect Flow (By taking water in first impinger) and Direct Flow (Without Water in First impinger). Similarly all experiments were conducted at different temperatures and at different times of interval for reaction. Operating condition of SO_2 absorption is given in table – 1. Experimental set up shown in figure – 12 and schematic diagram of experimental protocol shows in figure – 11

S. No.	Operating Condition	Value
1	Initial Concentration of Scrubbing media	Varying
2	pH of solution	Varying
3	Total liquid hold up	100 ml
4	Temperature of solution	Varying
5	Time period for reaction	Varying
6	Flow of flue gas in impinger	3 LPM
7	SO_2 load in flue gas	3000 – 3200 ppm
8	Flue gas Temperature	135 ℃
9	Flue gas flow in duct of ESP O/L	150522 M^3/hr
10	Pet Coke Feeding Rate	13 Ton/ hr
11	Lime Stone Feeding Rate	1.0 Ton/hr

Table 1. Operating condition of SO_2 absorption in Scrubbing media.

S. No	Concentration of Sludge Sample (%)	Initial Concentration of SO$_2$ at I/L of absorbing media (ppm)	Concentration of SO$_2$ at O/L of absorbing media (ppm)	Recovery of SO$_2$ (%)
1	5.00%	2950	1134	62.56
2	10.00%	2950	1205	60.18
3	15.00%	2950	1444	52.08
4	20.00%	2950	1734	42.25
5	25.00%	2950	1795	40.16

Table 2. Effect of Conc. of sludge solution and recovery of SO$_2$.

S. No	Concentration of Sludge Sample (%)	Yield of precipitate (g)	Mg^{+2}	Percent CaSO$_4$	Percent SO$_2$ (By Volumetric)	Percent Alkalinity	L/G ratio
1	5.00%	5.25	7.72	5.54	2.6	0.0014	33.89
2	10.00%	10.56	5.91	5.20	2.44	0.0028	50.84
3	15.00%	15.56	4.34	3.65	1.72	0.0144	67.79
4	20.00%	20.56	2.55	3.09	1.45	0.0158	84.74
5	25.00%	25.89	1.95	1.73	0.81	0.0201	101.69

Table 3. Analysis results of precipitate which was prepared by sludge solution and SO$_2$.

S. No	Concentration of Ca(OH)₂ (%)	Initial Concentration of SO₂ at I/L of absorbing media (ppm)	Concentration of SO₂ at O/L of absorbing media (ppm)	Recovery of SO₂ (%)
1	5.00%	2980	621	80.15
2	10.00%	2980	739	76.2
3	15.00%	2980	898	70.85
4	20.00%	2980	1097	64.18
5	25 .00%	2980	1279	58.06

Table 4. Effect of Conc. of $Ca(OH)_2$ solution and recovery of SO_2.

S. No.	Concentration of Ca(OH)₂ Sample	Yield of precipitate (g)	Mg⁺²	Percent CaSO₄	Percent SO₂ (By Volumetric)	Alkalinity Percent	L/G ratio
1	5.00%	5.55	3.61	8.75	4.12	0.02	33.55
2	10.00%	10.89	2.41	7.51	3.53	.0216	50.33
3	15.00%	15.06	1.96	7.34	3.45	.0252	67.11
4	20.00%	20.42	1.29	5.03	2.36	.0324	83.89
5	25.00 %	25.18	1.24	4.18	1.96	.0540	100.67

Table 5. Analysis results of precipitate which was prepared by $Ca(OH)_2$ solution and SO_2.

S.No.	Concentration of NaOH (%)	Initial Concentration of SO₂ at I/L of absorbing media (ppm)	Concentration of SO₂ at O/L of absorbing media (ppm)	Recovery of SO₂ (%)
1	5 %	3067	75	97.96
2	10 %	3067	158	95.08
3	15 %	3067	306	90.18
4	20 %	3067	324	88.02
5	25 %	3067	455	85.19

Table 6. Effect of Conc. of NaOH solution and recovery of SO_2.

S. No.	Concentration of NaOH Sample	Yield of precipitate (g)	Percent SO₃ (By gravimetric method)	Percent Na₂SO₄	Percent SO₂ (By volumetric)	Percent Alkalinity	L/G ratio
1	5 %	4.88	20.76	35.49	39.21	0.62	16.30
2	10 %	9.76	5.67	17.00	25.61	1.17	32.60
3	15 %	14.15	1.49	9.81	20.54	1.64	48.90
4	20 %	18.62	0.52	5.77	19.47	1.68	65.21
5	25 %	23.28	0.24	3.99	17.62	1.75	81.51

Table 7. Analysis results of precipitate which was prepared by NaOH Solution and SO₂.

S. No.	Time (Min.)	pH of solution	Volume of 1 M Oxalic acid consumed in titration using phenolphthalein indicator (ml)	Conc. of NaOH (%)
1	0	12.57	20.05	80.06
2	15	10.62	15.56	62.2
3	30	8.82	3.5	14.2
4	45	7.95	1.23	4.8
5	60	5.62	0.56	2.2
6	75	4.75	0.32	1.2

Table 8. Effect of pH of NaOH solution for absorption of SO₂.

S. No.	Flow of SO₂ gas	Initial Concentration of SO₂ (ppm)	Concentration of SO₂ after formation of Sodium sulphate (ppm)	Recovery (%)
1	Direct	3050	145	95.25
2	Indirect	3050	1818	59.62

Table 9. Effect of direct and indirect flow of flue gases in NaOH solution and removal efficiency of SO₂.

S. No.	Flow of SO_2 gas	Yield of precipitate (in gm)	% SO_3 (Gravimetric)	% Na_2SO_4	% SO_2 (Volumetric)	Alkalinity (%)
1	Direct	9.55	1.79	3.17	39.21	1.12
2	Indirect	9.02	.233	.413	20.01	1.72

Table 10. Analysis results of precipitate which was prepared by varying the flow of flue gases in NaOH solution.

S. No.	Temperature of NaOH solution	Initial Conc. of SO_2(ppm)	Conc. of SO_2 after formation of Sulphate(ppm)	Recovery (%)
1	20-25 ∘C	3080	302	90.18
2	25-30 ∘C	3080	566	81.62
3	30-35 ∘C	3080	675	78.08

Table 11. Effect of temperature of NaOH solution and recovery of SO_2.

S.No	Temperature of NaOH Solution	Yield (g)	% SO_3	% SO_2	% Na_2SO_4	% Alkalinity
1	20-25∘C	9.77	0.62	38.72	1.100	1.68
2	25-30∘C	9.25	0.42	31.92	0.745	1.80
3	30-35∘C	9.06	0.22	17.87	0.390	1.95

Table 12. Analysis results of precipitate which was prepared by different temperature of NaOH solution and SO_2.

S.No	Time for reaction(Min)	Initial conc. of SO_2 (ppm)	Conc. of SO_2 after formation of SO_4(ppm)	Recovery (%)
1	20	3075	761	75.25
2	40	3075	609	80.18
3	60	3075	360	88.27

Table 13. Effect of time intervals of reaction and recovery of SO_2.

S.No	Time of Reaction (Min)	Yield (g.)	% SO_3	% SO_2	% Na_2SO_4	% Alkalinity
1	20	8.62	1.63	26.68	2.89	1.16
2	40	8.95	2.35	29.34	4.17	0.96
3	60	9.02	3.06	36.51	5.03	0.75

Table 14. Analysis results of precipitate which was prepared by different times of intervals of reaction between NaOH solution and SO_2.

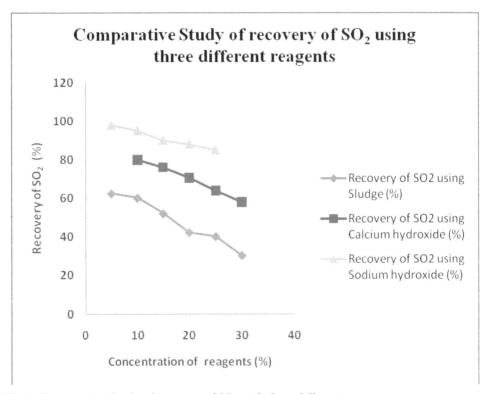

Fig. 1. Comparative Study of recovery of SO_2 with three different reagents.

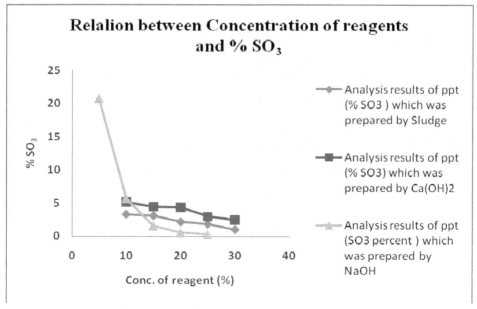

Fig. 2. Comparative Study of Conc. of three different reagents with % SO₃ (Gravimetric) of precipitate.

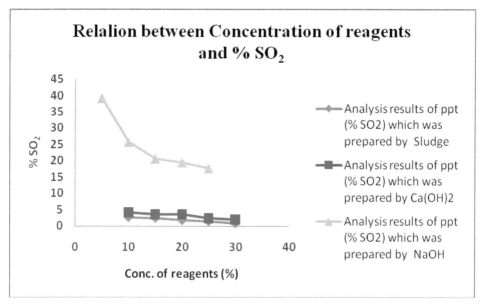

Fig. 3.Comparative Study of Conc. of three different reagents with % SO₂ (Volumetric) of precipitate.

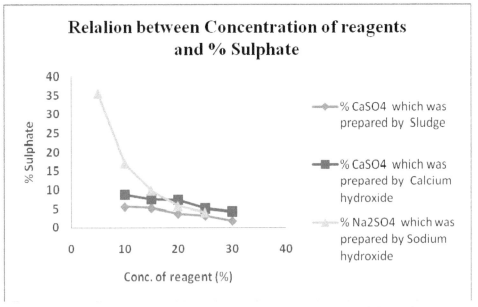

Fig. 4.Comparative Study of Conc. of three different reagents with % sulphate of precipitate.

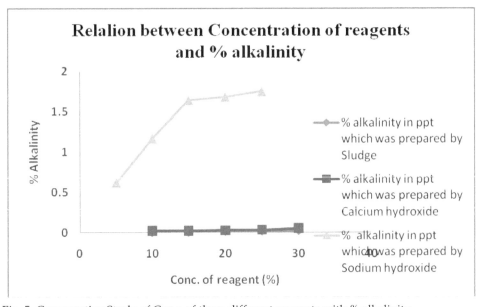

Fig. 5. Comparative Study of Conc. of three different reagents with % alkalinity.

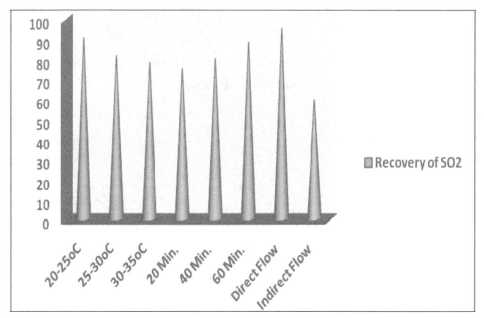

Fig. 6. Comparative Study of recovery of SO₂ with different parameters.

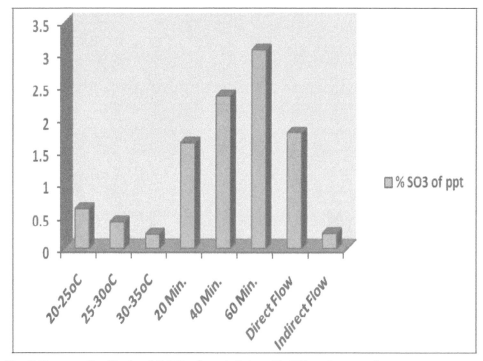

Fig. 7. Comparative Study of % SO₃ of precipitate with different parameters.

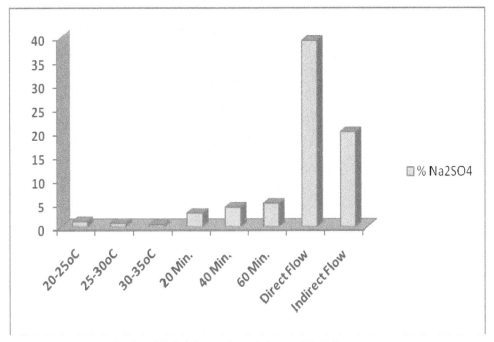

Fig. 8. Comparative Study of % Sulphate of precipitate with different parameter.

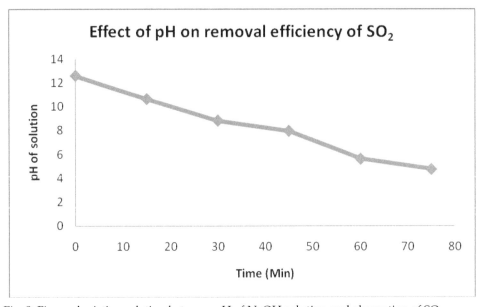

Fig. 9. Figure depicting relation between pH of NaOH solution and absorption of SO₂.

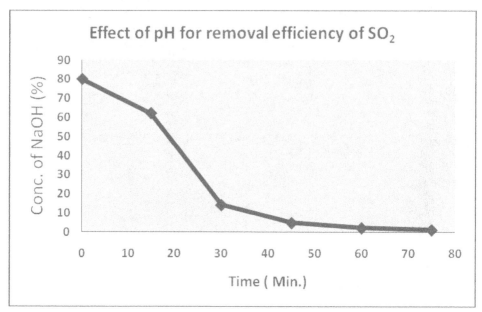

Fig. 10. Figure depicting relation between time period and falls in conc. of NaOH.

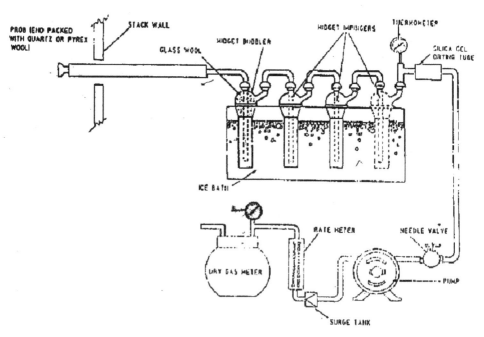

Fig. 11. Schematic diagram of experimental protocol.

Fig. 12. Experimental Set Up by using SO_2 monitoring kit for absorption of SO_2.

Fig. 13. Experimental Set Up by research scholar using SO_2 monitoring kit.

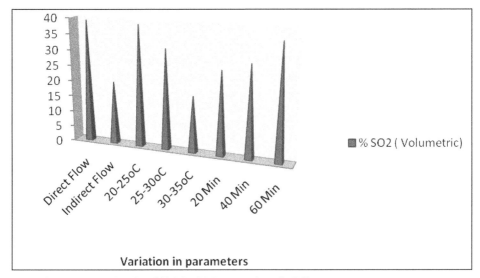

Fig. 14. Comparative Study of % SO₂ (Volumetric) with different parameters.

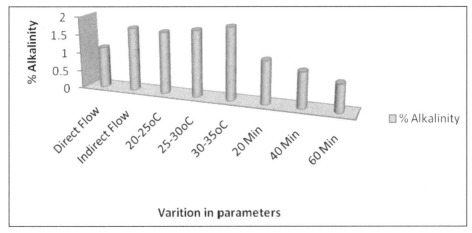

Fig. 15. Comparative Study of % Alkalinity with different parameters.

5. Result and discussion

Table -2 to 7 reports that relation between recovery of absorption of SO₂ using varying concentration of Sodium hydroxide, Calcium hydroxide, and Sludge with analysis results of precipitate. As can be seen from figure – 1 that recovery of SO₂ using, Calcium hydroxide, and Sludge is far below that using Sodium hydroxide. Figure -2 shows the results of % SO₃ (Gravimetric) of precipitate which was prepared by three different reagents and SO₂ contained in flue gases. It is reported that % SO₃ is higher in case of NaOH as to others. Figure -3 shows the results of % SO₂ (Volumetric) of precipitate which was prepared by three different reagents and SO₂ contained in flue gases. It is reported that % SO₂ is higher in case of NaOH as to others. Figure -4 shows the results of % respective Sulphate of precipitate which was prepared by three different reagents and % SO_2 in flue gases. Figure - 5 shows the results of % alkalinity of precipitate which was prepared by three different reagents. We know that alkalinity is the reverse of % SO_2 and it is confirmed by figure – 5. Figure -9 and table – 8 reports that effect of pH of NaOH solution and absorption of SO_2 and it is confirmed that when increase in the time period for absorption of SO_2 in NaOH solution, then there is a significant decrease in pH. Figure -10 reports that with the increase of time period for absorption of SO_2 in NaOH solution there is a significant decrease in conc. of NaOH Solution. Table -9 shows recovery of SO_2 using different parameters like time period for reaction , temperature of Solution and flow of flue gases in impingers with analysis results of precipitate. Figure – 6 reports that recovery of SO_2 with different parameters. Figure – 7 reports that % SO_3 in precipitate which was prepared by exhaust SO_2 using different parameters. Figure – 8 reports that amount of % Sulphate which was prepared by SO_2 using different parameters. Figure – 14 reports that amount of % SO_2 (Volumetric) which were prepared by exhaust SO_2 using different parameters. Figure – 8 reports that amount of % Alkalinity which were prepared by SO_2 using different parameters.

6. Conclusion

From the comparative study of three different reagent regarding to removal of SO_2 , it is observed that Sodium hydroxide is superior as compare to calcium hydroxide and sludge. The initial rate of absorption is higher for Sodium hydroxide as compared to calcium hydroxide and Sludge. All the absorption methods coupled with a chemical reaction. It may be suggested that Sulphur dioxide is a weak acid and it is a well known fact that reaction of a weak acid with a strong base is fast, meaning stronger the base faster would be the reaction Therefore Sodium hydroxide is a strong base compared to calcium hydroxide and sludge so this evident that Sodium hydroxide is a better solvent for removal of SO_2.

The lower Conc. of the reagent is found to be optimum. Increasing conc. of solution is not very fruitful for maximum absorption of SO_2 in exhaust flue gases. This is because of load of SO_2 in flue gases is very low (at ppm level), so the reagent remains as it is in solution after completely absorption of SO_2.

The pH of the solution should be alkaline. Because of nature of SO_2 is acidic and reaction is restricted in acidic solutions

The temperature of solution should be lower i.e. 20- 25 ºC. Because of at higher temperature reversible reaction may be take place and partially formed product may be change in to initial reactants.

The time period of the absorption of SO_2 should be maximum for completely absorption of SO_2.

The direct flow of flue gases in to impingers containing solution will results maximum absorption of SO_2 instead of indirect flow of flue gases because of in indirect SO_2 react with water form sulfurous acid.

On the basis of our study we can recommended that if flue gas desulphurization system (FGD System) is set up before Chimney then maximum SO_2 is trapped, resulting lowers the SO_2 conc. in environment and lowers the air pollution.

7. FGD design and equipment

7.1 Purpose

Air pollution is one of the very important issues world-wide. The wet limestone-gypsum process has been the most popular method adopted to eliminate SO_2 emitted from thermal Power Stations. However, due to the relatively high construction cost, its further implementation has inevitably limited and the development of more economical FGD technology has been sought.

Hence, Hitachi Compact FGD System was developed, for the purposes of simplification and cost reduction utilizing features of the latest FGD technology fully. The first System was delivered to Peoples Republic of China under the"Green Aid Plan", which has been organized and managed by the Ministry of International Trade and Industry (MITI), Japan, in order to implement their policy to transfer environmental preservation technology to neighbouring countries and it contributes to global environmental preservation and the technologies, such as, higher gas velocity in the absorber and adoption of horizontal flow spray absorber instead of conventional vertical flow spray absorber shortened duct length. Eventually it helps to accomplish a lower construction cost.

7.2 Performance

The absorption and forced oxidation mechanisms are the same as the conventional wet limestone gypsum FGD technology, so it is possible to achieve more than 80% of SO_2 removal efficiency. Also, because of higher gas velocity under the same conditions of gas versus liquid ratio, it is possible to maintain the same SO_2 removal efficiency.

In the method of horizontal spray tower, it is possible to achieve high dust removal efficiency as in the vertical spray tower.

7.3 Special features

1. Absorber, having functions of dust removing, SO_2 absorbing, SO_2 oxidization simultaneously
2. Adoption of horizontal flow spray absorber and simplified flue duct make a compact arrangement.
3. Use of limestone of easy handling and low cost
4. High Ca utilization factor
5. Complete oxidation in the absorber
6. High dust removal efficiency at spray part
7. Re-use of by-product gypsum as salable gypsum
8. Easy to retrofit to existing plant

7.4 Process description

The SO_2 contained flue gas flow into FGD system through duct. The SO_2 is absorbed and removed by the chemical reaction of limestone slurry sprayed through horizontal flow spray as an absorbent. By injecting air in the absorber tank, the absorbed and removed SO_2 forms gypsum, then, the gypsum slurry is delivered outside system, as in the method of latest wet limestone gypsum FGD system.

In the absorber :

$$SO_2 \quad + \quad H_2O \quad \longrightarrow \quad H_2SO_3$$

$$CaCO_3 + 2H_2SO_3 \quad \longrightarrow \quad Ca(HSO_3)_2 \ + \ CO_2 \ + \ H_2O$$

In the tank :

$$Ca(HSO_3)_2 \ + \ O_2 + 2H_2O \quad \longrightarrow \quad CaSO_4.2H_2O \ + \ H_2SO_4$$

$$CaCO_3 + \ H_2SO_4 + H_2O \quad \longrightarrow \quad CaSO_4.2H_2O \ + \ CO_2$$

It shows the same reaction characteristics as in conventional one. From the absorbing tower's structural view point, adoption of higher gas velocity and horizontal spray tower eliminate certain portion of duct for up and down and both of Duct's capacity and system's cost can be reduced tremendously.

Also, it is possible to reduce both auxiliaries and installation space by making the system compact.

7.5 Process flow sheet

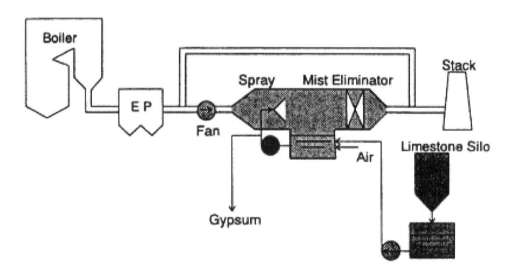

7.6 Outline of absorber

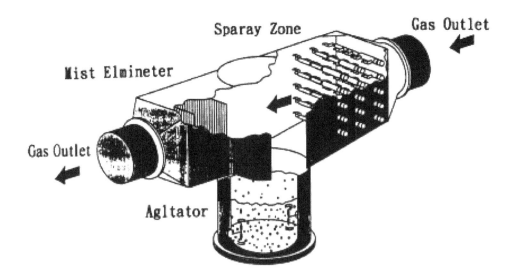

8. References

[1] Sheth K.N. Patel and Patel Neha J, effect of concentration in absorption of Sulphur dioxide with sodium hydroxide, *Env. Poll. Cont. J.*, 9, 14-18. (2006)

[2] B. Buecker, Wet lime stone Scrubbing fundamentals, *power engineering*, 110, 32-37, (2006)

[3] Ram S. Gosavi, Flue gas desulphurization – An Overview of system and technologies, *Env. Poll. Cont. J.*, 8 , 5-8. (2005)

[4] Maohong fan, C brown, a process for synthesizing polymeric ferric sulfate using sulphur di oxide from coal combustion , *Int. J. Env. Poll.*, 17 ,102-109. (2002)

[5] Shih-Wu Sung, Robert C. brown, synthesis, characterization and coagulation performance of polymeric ferric sulphate, *Int. J. Env. Poll.*, 128, 483-490. (2002)

[6] Aron D. Butler, Maohong Fan , Robert C. brown, Comparison of polymeric and conventional coagulants in arsenic removal, *J. Env. Engg.*, 74, 308-313, (2002)

[7] C. Dene, *FGD chemistry and analytical methods handbook* , power engineering, 2, 48-55, (2002)

[8] B. Buecker, gypsum seed recycle in lime stone scrubber, *power engineering*, 110, 10-12, (1986)

[9] Arthur kohl , Richard B. Nielsen, gas purification , *fifth edition gulf publishing company*, 10, 42-49, (1975)

[10] Slack A. V, Hollindon G.A, Sulphur di oxide removal from stake gases, *Second edition Noyes Data Corporation*, 11, 45-55 (1971)

[11] Indian standard method for *Measurement of emission from stationary sources.* IS 11255 (Part 2) 1985 IS 5182 (Part 2) 2001

[12] Indian standard methods of *sampling and test for quicklime and hydrated lime (1 st revision)* IS No. 1514.

[13] Indian standard methods of *chemical analysis of limestone , dolomite and allied materials ,determination of iron oxide , calcium oxide and magnesia (1 st revision)* IS No 1760 Part 3 1992.

[14] Indian standard methods of *Sodium bi Sulphite technical (fourth revision).* IS 248: 1987

[15] Indian standard methods of *Caustic soda, pure and technical (third revision).* IS 252 :1991

[16] Indian standard methods of *sampling and test (physical and chemical) for water used in industries:* IS 3025-1964.

9

Allergens, Air Pollutants and Immune System Function in the Era of Global Warming

Barbara Majkowska-Wojciechowska and Marek L. Kowalski

Departament of Immunology, Rheumatology and Allergy, Medical University of Łódź
Poland

1. Introduction

Nowadays, almost half of the world's population lives in or near areas where the quality of air is poor. Rapid changes in the environment related to the "industrial revolution" have changed the earth within almost no time. These civilization changes are usually positive, as they lead to technological advancement, raising living standards, providing better health care and hygiene. They are however accompanied by unforeseeable climatic changes and negative health effects, which result from progressive contamination of the biosphere, as well as excessive exposure to toxins and allergens. They, in turn, lead to a growing number of immune system dysfunctions or even deaths in animals and humans.

It is estimated that in the 20th century, only around 100,000 chemical substances were introduced. However, the pace at which new ones are being introduced is still growing. In 2009, 50 million synthesized chemical compounds were entered into the international register, "The world's largest substance database" and of that number, 143,000 compounds were used for industrial purposes. At present, as it has been recently updated (04 November 2011), the number of the registered compounds is now 63, 839, 600 which means an increase of more than 20% within less than two years [59]. It is a fact that the air in urban city agglomerations is contaminated with hazardous compounds which appear as a result of the process of incomplete burning of minerals used in transport, power and other industries. Some of them exacerbate the greenhouse effect and cause disorders in the immune system. Additionally, we should not neglect the widespread pollution of rural areas. Pollution also brings about disorders in the immune system in humans, domestic animals as well as wild animals. Although the concentration of certain chemical substances is monitored in the air, water and food, some studies have shown that the influence of the concentration on dysfunctions of immune system in an individual is very complex and often difficult to explain. When the homeostasis of the immune system is disturbed it is difficult to treat such cases.

The immune system is a highly complicated, even intelligent, system with a characteristic hierarchy. Studies show that the system often has difficulty in adapting to environmental changes. More and more immune pathologies are appearing. They start with immunosupression, immunomodulation and finish with autoagression and allergy. The reasons for these phenomena are still being examined. Epidemiological, clinical, as well as toxicological, examinations conducted on humans and animals are the grounds for evaluation of health effects resulting from environmental pollution. The analysis of factors related to local and global environmental changes, particularly the toxicology of the urban environment, which might affect the proper functioning of the immune system is highly important from the point of view of public health. It is known already that environmental toxins can lead to reduced resistance of the epithelium and facilitate the contact of inhalatory allergens with the "network" of dendritic cells. The activation of these cells by toxic compounds make the allergens more recognizable. The exposure to the factors might lead to oxidative stress through the production of reactive oxygen (ROS) and nitrogen species (RNS) at an amount which cell defensive mechanisms cannot manage.

Environmental toxins, depending on the dose, can cause permanent changes, including epigenetic ones. Recently, the clinical evaluation of immune system dysfunctions has been thoroughly analysed; the genetic and environmental aspects are taken into consideration, as well as any climatic changes which might lead to an increase in the allergogenic potential of natural organic compounds that might directly affect the immune system [98]. All these factors and related compounds are potent. Their health implications might be unforeseeable, instant, long-term, irreversible or permanent. The aim of the study is a review of modern environmental toxins, their relationship to the process of global warming, the appearance of disturbances in the immune system as well as chronic inflammatory diseases of the respiratory system.

2. Classification of air pollutants

Air pollutants are usually classified as natural and anthropogenic. Another classification also takes the environment into consideration.

2.1 Natural pollutants

Natural pollutants – geological (e.g. volcanic ash emission, fires, emission of minerals coming from soil erosion). On the European scale, the pollution is monitored with satellite data from the UE MACC 2010 international research project. The European database for biogenic pollution (from 35 countries in Europe): pollen and fungal spores, can be found on http://www.polleninfo.org/.

2.2 Anthropogenic pollutants

Anthropogenic pollutants – chemical compounds which are produced as a result of human activity. They appear in industrial and metallurgical technological processes; they are combustion products of fossil fuels, dust and aerosols. The last category comprises pollutants which pose a health hazard for a man - NO_x O_3, SO_2, CO, nicotine smoke, hydrocarbons, lead, chromium, particulate matter (PM), including nanoparticles. There has been a rapid increase in the number of reports demonstrating a close relationship between the exposure to anthropogenic substances and disorders of the immune, respiratory and cardiovascular systems or even a growing number of deaths. Findings show that poor air quality results in dysfunctions of the immune system. They, in turn, lead to the poverty of people inhabiting the particular region [105,118]. The classification of pollutants is also based on the number of particles in the air, their origin and state of aggregation, while a further classification concerns the place where the pollutants are measured e.g. city/rural environment, air inside/outside a building. This study aims at presenting the influence of particular toxins on human health and we try to give a detailed description of those toxins one by one.

2.3 Dirty dozen: Examples of compounds which heavily pollute the environment and cause immune pathologies

The Environmental Protection Act and directives issued by the European Parliament and the European Council as of 21 May 2008 provide the necessity of constant monitoring of air and 12 of its components which pose a health hazard:

- nitrogen dioxide NO_2,
- sulfur dioxide SO_2
- carbon monoxide CO,
- benzene C_6H_6,
- ozone O_3,
- particulate matter PM_{10},
- lead Pb in PM_{10}
- arsenic As in PM_{10},
- cadmium Cd in PM_{10},
- nickel Ni in PM_{10},
- benzopyrene in PM_{10},
- particulate matter $PM_{2.5}$.

According to the provisions of the European Union, this act must enforced in all member states and its established regulations are required to be respected by cities with populations above 100,000 [37].

2.4 Urban outdoor pollution

Large cities are spread all over the world. In Europe, around 75% of the population live in cities. Therefore, the majority of scientific research on the topic of air pollution is performed in cities. The pace of migration from rural areas to cities is still fast. Unfortunately, city dwellers pay a huge price for living in large agglomerations where even breathing puts their life at risk. The air in major cities contains over 700 substances, both organic and inorganic compounds, as well as metals (e.g. lead, cadmium, manganese, mercury) and their compounds with other stable inorganic and organic substances, including DEP, soot, dust, smoke, dioxins, metals, polycyclic aromatic hydrocarbons (PAHs) – pyrenes, benzoapyrenes, dibenzofurans, polychloro aromatic hydrocarbons (VOC), fungal and pollen allergens and many others.

In polluted city environments, there is a deficiency of bacterial flora with probiotic and saprophytic properties (e.g. bifidobacterium, lactobacillus) and so the inhabitants are exposed to pathogenic microorganisms (e.g. Staphylococcus ureus, Candida albicans) [86]. It has also been confirmed that dust collected in city areas is characterized by higher oxidative activity, mainly because of the presence of metal ions Fe^{3+}. In healthy people a typical concentration of dust (< 0.2 ppm) usually causes reversible effects such as irritation of the mucous membrane in the conjunctiva and upper respiratory tract.

2.5 Rural outdoor pollution

Air quality in rural areas is much less frequently monitored. Intensive agricultural development and other industries contribute to progressing soil and water pollution with the remnants of fertilizers and pesticides, which impoverishes soil and causes wood atrophy (Figure 1). In water reservoirs, algae bloom, block access to oxygen and cause eutrophication

Fot. Agnieszka Wojciechowska

Fig. 1. The effects of plant vegetation in unpolluted (Łódź, Arurówek) and polluted (Krzewo) areas in central Poland.

of water ecosystems. The agricultural industry also contributes to the excessive emission of nitrogen compounds; NH_3 and NH_4^+ ions have been used on a huge scale in artificial fertilizers since the 1st half of the 20th century. Nitrogen compounds and other substances coming from rural areas are bound by PM particles and often transported over longer distances by acid rain. Many compounds which are found in weed-killing agents may irritate the respiratory system and even block receptors for hormones or neurotransmitters; many pesticides are inhibitors of acetylcholinesterase. They negatively influence the homeostasis of the body and its immune system. Pesticides and toxic herbicides may also have immunotoxic and immunosuppressive properties. They can act by causing deactivation of lysozyme, an important protective factor against pathogenic Gram(+) bacteria present in saliva, tears, secretion of mucosa and body fluids in phagocytes [20]. On the level of adapted immunity, they can act by, among other things, inhibiting lines of T CD4(+), CD25(+)lymphocytes and T reg lymphocytes [30].

2.6 Indoor pollution

Poor air quality inside a building might constitute a serious environmental threat [41]. It has been confirmed that air pollution inside buildings is usually 2 - 5 or even 100 times higher than outside [62]. Interestingly, not only pollutants coming from outside are responsible; pollen, house-dust allergens such as mite allergens and cockroach allergens, as well as fungal spores, animals, household cleaning products, building materials, air fresheners, naphthaline and even dry-cleaned clothes also exert an allergenic effect [40]. House owners are also threatened with breathing fumes emitted during the process of burning traditional biofuels, such as gas, coal and wood, as well as highly toxic nicotine smoke, which contains more than 6000 toxins. One such component of cigarette smoke is nitrogen oxides, whose level ranges between 200 mg/m^3 and 650 mg/m^3 [147]. The WHO defines the 1-h limit value of 200µg NO_2/m^3 air. Due to their poor water solubility, the site of immunotoxicity of nitric oxides is the upper respiratory tract. The unfavorable situation inside buildings could be improved quite easily. For example, the use of chemical products could be reduced or the building occupants could stop smoking nicotine, since the filtration of air polluted with nicotine smoke is hardly effective [22].

Another world concern is pollution with bisphenol or tributyltin (TBT), compounds which can be found in house dust, as well as outside, and which can seriously impair the immune system in humans and animals.

Bisphenols (PBA, PBC) belong to a group of extremely dangerous toxic indoor compounds which have been in use for more than 50 years. Bisphenols can be found in many everyday products made of polycarbonate. Exposure to bisphenols is hazardous even in intrauterine life. Bisphenols have been identified in the blood of infants whose mothers used simple devices for recycling domestic waste while pregnant; they were exposed to polychlorinated biphenols (PBC) by breathing them in. The infants' birth weight was lower and their Apgar scores were worse [146]. The list of sources of biphenols is long. The compounds accumulate in house dust, which is highly dangerous for children. They can be found in commonly used plastic objects such as bottles for babies (although the production of bottles containing BPA in EU countries was banned in March 2011, and the import and sale of such bottles was banned in June 2011), plastic bottles for milk and drinks, kitchen utensils, wrapping food foil, dental fillings, dentures, dental bridges, food and medicine packaging, as well as plastic elements within various medical devices such as inhalers, dialyzers, pacemakers, contact lenses and protective masks. They are also components of paper and are found in fire retardant agents in furniture and fabrics as well as in electric cables.

Bisphenol easily penetrates the skin, air passages and the alimentary tract. It can induce dangerous epigenetic changes in reproductive cells, disturbing chromosome division (aneuplodia) thus potentially leading to Down syndrome. After heating or sterilizing these articles (e.g. bottles for babies) polycarbonates are slowly degraded and, as a consequence, bisphenol molecues are released dozens of times faster. Due to the similarity of PBA molecues to estrogens, these compounds induce many hormonal disturbances, mainly in the thyroid and pituitary gland; they also infertility, neoplastic diseases , diabetes, obesity as well as have neurotoxic properties [125].

Due to their high toxicity, tributyltins (TBT) are also regarded as some of the most dangerous substances polluting the indoor environment. TBT have been found in the air inside many public buildings and private houses [97]. Many analyses have confirmed their immunosuppressive character, which might lead to impairment of anti-infectious immunity and development of neoplasms. Despite this, TBT have been widely used in agriculture and industry since the 1960s. Nowadays they can be found in plastics, herbicides, household cleaning agents and building materials despite a considerable body of alarming information concerning their negative effects. They are in recycled products and their production has not been banned, yet. They are used for conserving wooden objects, hulls in ships, in textiles, plastic products, fungicidal, insecticidal and bactericidal agents, commercial catalysts and even toiletries, especially in antibacterial additives for washing liquids and soaps, pampers diapers and toys for children [74]. These toxins have also been traced in many alimentary products, even in the milk of breastfeeding mothers. Toxicologists warn against using plastic boxes, toys and other products, even those which were permitted to be sold by the FDA, because of their potential effect on cell metabolism. By disturbing endogenic steroid hormones and modifying the activity of many drugs, they induce irritation of respiratory passages, skin and eyes, and have neurotoxic properties [133]. It has been confirmed that TBTs activate peroxisome proliferator-activated receptors (PPARs), present on cells, which might lead to metabolic disorders, upset the lipid balance and as a consequence, lead to diabetes and obesity [29]. Nevertheless, there are hopes that recently isolated bacterial strains (class Bacilli and c-Proteobacteria) might play a role in the biodegradation of TBT [4].

2.7 Air Quality Index (AQI)

In most countries all over the world, the level of air pollution is constantly monitored. However, for many people, the data is inaccessible or cannot be understood. In American countries there is one general air pollution indicator called the Air Quality Index (AQI) [60], thanks to which, it is possible to provide a daily practical and yet complex evaluation of many pollutants on the American continent and show to what extent air quality might have a negative influence on health. The AQI values are presented on a six-grade scale. In European countries there are initiatives to introduce a uniform index evaluating air quality which might improve the system of sending information and warnings. The levels of EPA Air Quality Index and their relationship with oxidative stress biomarkers in epithelial cells are shown in figure 2.

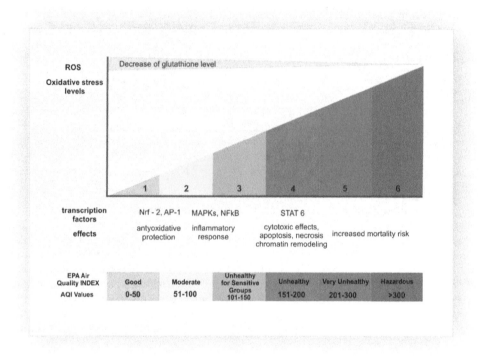

Fig. 2. Levels of oxidative stress biomarkers in epithelial cells and their relationship with the EPA Air Quality Index. It has been developed on the basis of the "Air Quality Index (AQI) - A Guide to Air Quality and Your Health" http://www.airnow.gov/index.cfm?action=aqibasics.aqi, and Xiao GG, Wang M, Li N, Loo JA, Nel AE. Use of proteomics to demonstrate a hierarchical oxidative stress response to diesel exhaust particle chemicals in a macrophage cell line. J Biol Chem. 2003;12;278:50781-90.

2.8 Allergens and environmental toxins

A great number of atmospheric aerosols are of biological character. It has been estimated that bioaerosol constitutes 5 – 10% of the total volume of PM [100]. Plant pollen and fungal spores are dominant in rural areas rather than in cities. However, city inhabitants more frequently suffer from pollen allergy [95]. According to scientific findings, the main factor favouring pollen allergy might be oxidative stress connected with the synthesis of NADPH endogenous oxidase in pollen grains such as those of birch or ambrosia, and air pollution of oxidative character. They can have a negative, pro-oxidative influence on cells of the epithelium of the respiratory tract, activation of dendritic cells and stimulation of pro-inflammatory processes which might contribute to the development of the allergy [113,5]. The evaluation of the dose-response relationship between the exposure to pollen and allergy symptoms is unclear. An interesting study was performed by scientists in Basel, Switzerland in which they gathered a database on the levels of birch and grass as well as the incidence of allergy over a period of 39 years. It concluded that within 20 years (from 1969 to 1990) with the increase in the level of pollen, there was an increase in the incidence of hay fever and allergy to birch and grass pollens. In the two subsequent decades (from 1991 to 2007) scientists observed the decrease in pollen of these taxons, which was accompanied by a decrease in the incidence of allergy [45]. Other experiments have not shown a dose-response relationship between the sIgE level in serum and exacerbation of clinical symptoms induced by Timothy pollen [65]. It should be remembered that environmental allergens are most often the proteins of living organisms sensitive to changes in the natural environment, such as temperature, osmolarity, humidity, pH and exposure to toxins. Modern methods of molecular research facilitate introducing detailed patterns of environmental stress response (ESR) caused by environmental pollution; not only in terms of genetics, proteomics and kinetics of their synthesis but also in terms of the virulence/allergenicity of proteins, as well as the response of the immune system to synthesized compounds. The synthesis of co called pathogenesis-related proteins (PRS), which might be both allergens and toxins, is one of the forms of plant and fungal spore response to air pollution.

2.9 Pollen allergens and environmental pollutants

Plants are considered to be highly efficient, ecological air filters whose filtering capacity grows with the growing level of pollutants. They are able to absorb and accumulate magnetic nanomolecules (Fe_3O_4) [151] and even 97% of toxic VOCs (volatile organic compounds) from the atmosphere [72]. Rarely are the potential consequences for flora, fauna and a human considered. We already know that air pollution absorbed from the atmosphere contributes to the growth, development and yielding of plants. They have a toxic influence on generative tissues and decrease the production of plant pollens; for example, herbicides commonly used for killing weeds inhibit the production of pollens in arable crop [36].

Pollen grains are too large to get to the lower airways and the symptoms of allergic reaction depend on allergens emitted after the destruction of the pollen exine. It has been concluded that gas pollutants, found in PM, react with the hydrocarbons of the cell wall leading to the modification of its structure and changes in the shape of the pollen grain, which facilitates the release of allergenic proteins. Sometimes, under the influence of these changes, plants produce a greater number of grains but of smaller size. They can also penetrate the plant tissues and interfere with their physiological processes, the course of photosynthesis and the cell metabolic pathway. They induce some damage to cells and mutations. For example, upon penetrating the cell, SO_2 can react with water and become sulfuric acid, which inhibits the process of photosynthesis and damages cells [115].

Pollen allergens can remain in the air for as long as for a few weeks and join many pollutants. The presence of aeroallergens/air pollutants interactions appears to increase the morbidity from aeroallergens, but immunomodulation mechanisms are not clear. It is possible that allergens interact with environmental pollutants to increase their interactive effects (figure 3). Microscopic analysis showed that in a polluted environment, the exine (outer layer of the pollen) thinner, distorted and more susceptible to breaks, which favours the emission of allergens [26]. Also allergens emitted from pollen grains can change their properties under the influence of chemical substances [115]. It was confirmed that O_3 induces changes in intracellular signal transduction, which leads to changes in gene expression and changes in the structure of allergen proteins. Air pollution might also induce post-translational modifications of allergenic proteins. Chehregani et al. [25] confirmed that after 5 to 10 days' exposure to Diesel's particles (DEP) there was an increase in allergenic potential of Lilium martagon pollen grains; immunoblotting subsequently confirmed the existence of modified or new protein (35 kD), which reacted with IgE antibodies of sensitized animals. Bommel at al. [18] stated that pyrene – an important component of DEP can aggravate an allergy by the induction of IL4 production. Exposure of human nasal epithelium to DEP and allergens (in vivo) contributed to a decrease in the IFNα and an increase in Th2 cytokines [135]. DEP particles can also act as allergic carriers, which intensify the presentation process by antigen presenting cells (APC) [66]. DEP particles can also induce epigenetic changes and in that way stimulate Th2 response, as confirmed by in vivo research [93]. The destruction of the exine by environmental toxins favours the release of

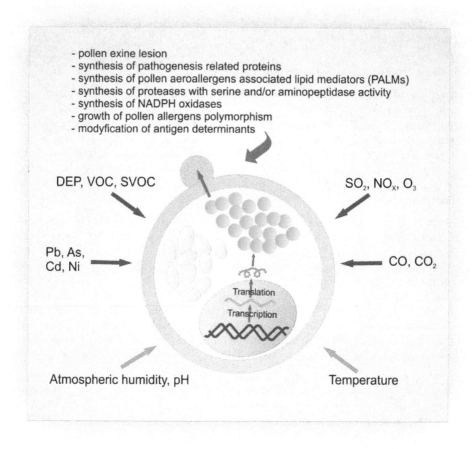

Fig. 3. Greater allergenicity of pollen allergens, collected from polluted areas, would be determined by chemical reactions between pollutants and pollen extracts and nonspecific modulation of synthesis machinery for many proteins, lipids and enzymes in response to environmental stress. Adopted from: Risse U, Tomczok J, Huss-Marp J, Darsow U, at al. Health- related interaction between airborne particulate matter and aeroallergens (pollen). Journal of Aerosol Sci. 2000;31, Suppl.:27-28.

phytoprostanes (lipophylic counterparts of prostaglandins) from pollen grains. The phytoprostanes are highly water soluble; they can get to the lower airways and inhibit the production of IL12 by dendritic cells, thus favouring Th2 response [31].

Many scientists point out the effect of pollution on climate warming, intense expansion of thermophilic plant species and readier exposure to their allergens, e.g. ambrosia allergens [152]. It is true that in some countries climatic warming lengthens the flowering season, contributes to the increase in the number of pollen grains, synthesis of modified proteins and allergenicity of some of the proteins. Strong winds resulting from weather anomalies transport pollens and their allergens. Together with toxins, which react with pollens and allergens, they cover long distances.

In their studies, Eckl-Dorna et al [38] showed that increased concentration of ozone contributes to the increase in the number of allergens of rye pollen from groups 1, 5, 6 as well as panallergens such as profilins. However, according to other studies, some air toxins can reduce the level of allergens. The exposure of grass pollen to an increased level of many toxins simultaneously, e.g. NO_2+O_3 or NO_2+SO_2 induced the decrease in levels of typical allergens; Phl p1b, Phl p4, Phl p5, Phl p6 and Phl p13, detected with an immunoblotting method and specific IgE antibodies of sensitized rats. The decrease in the levels of these allergens, according to the authors, might have resulted from mechanical or post-translational modifications of these proteins, under the influence of pollutants. It was difficult to detect them since the test highly sensitive and specific. There are plans to carry out further studies with the use of spectrometric methods which should make it easy to describe possible molecular changes in the epitopes of pollen allergens.

The existence of pollen grains in the atmosphere is based on the regular cycle of vegetation and flowering. The exposure to plant pollen fungal spores depends on many factors, among others, geographical latitude, global and local air circulation, weather factors, the level of urbanization of the environment and air pollutants. Numerous parameters characterizing climate warming, perceivable in many regions all over the world, influence the course of phenology and pollen distribution. Direct factors, such as temperature and precipitationas, well as more extreme weather phenomena, either induce the production of pollen or reduce it.

The indirect effects of higher temperature include the early blooming of trees, transport of pollen over a long distance and modification of flora (including the expansion of thermophilic taxons of plants such as ambrosia). In many centres which carry out long-term agrobiological and phenological studies, earlier and longer flowering seasons of allergenic plants have been observed. Biological observations performed by Ariano et al. [9] in Italy for 27 years showed a 25% increase in the level of pollen as well as an increase in the length of pollen seasons Parietaria (=85 days), olive (=18 days) and cypress (=18 days). It was shown that intensive exposure to pollen allergens contributed to the increase in the incidence of allergic episodes in people sensitized to pollen allergy in comparison to mite allergens. Kosisky et al [81] published the results of their twelve-year observations from Washington. They stated that although the level of pollens measured globally did not increase, the number of days with higher pollen counts did: by about 258% for grasses and by almost 12% for trees. Studies conducted in Switzerland over 21 years, gathered by Clot

[28] showed no particular changes in the pollen level. It should be pointed out that intensive production of pollen by plants can depend on many other factors, such as artificial fertilizers in the environment, an increase in the size of rural areas because of not cultivating arable land, development of soil erosion and desertification [112].

Apart from plant pollen levels, aeroallergens are also the subject of studies. Singer et al [123] showed a dangerous increase in allergen content in ambrosia pollen grains under the influence of increased concentrations of greenhouse gases. The response to the raised amount of CO_2 was a raised level of the most important allergen of that taxon, Amb a1, despite the fact the number of ambrosia pollen grains produced did not increase [21]. The results of a study performed in London showed the effect of allergen activity and air pollution ($<PM_{10}$) synergy can be spectacular; asthma was aggravated as a result of a 25% increase in IgE levels in the serum of patients sensitized to grass or tree pollen [50]. Air pollutants induce adjuvant effects which favour inflammatory processes, immunization and aggravation of asthma and allergy. Many studies highlight a positive relationship between the number of hospitalizations, high level of plant pollen and exceeding the maximum concentration of air chemical pollutants, especially SO_2 and PM particles.

2.10 Environmental pollution and pathogenesis-related proteins (PR)

It was confirmed that in unfavourable environmental conditions such as a pest invasion, the presence of microorganisms and fungal spores, exposure to fertilizers, air pollution caused by ozone, ultraviolet rays, phytohormones, pesticides, artificial fertilizers, the growth of temperature, plant cells synthesize pathogenesis-related proteins (PR), which are very often allergenic for humans. Interestingly, about 25% of the described plant allergens gathered in the database of International Union of Immunological Societies (IUIS), are (pathogenesis-related (PR) proteins. The mechanisms of their synthesis are still unknown. PR proteins, including, among others, isoflavone reductases and plant antibiotics, are numerous and usually belong to various families. The most common PR proteins have been classified into 17 families, marked PR1 – PR 17 according to taxonomic classification, molecular sequence, biochemical and functional characteristic as well as the response generated by the human immune system [137,138].

Many PR proteins have a high allergenic potential and great cross-reactivity. For example, PR-3 is a chitinase, the synthesis of which is an common reaction of plants to heavy metals, and the expression of Cup a 3, a PR-5 cypress allergen, is significantly higher in the pollen of plants grown in polluted areas [127]. PR proteins from group10 include ribonucleases and proteins similar to the main allergen of birch pollen: Bet v 1 and e.g. hazelnuts: Cor a 1 [124]. The enhanced synthesis of PR in plant cells and pollen contributes to a growing number of allergic reactions, including dangerous oral allergy syndrome (OAS) and even analyphylactic reactions. In sensitized patients suffering from pollen-food allergy syndrome, the first syndrome is allergy to inhalant allergens [19], especially in the city environment.

Interestingly, until recently, pollen grains have been considered simple, even primitive structures, dependent on vegetative organs of plants and inactive in a metabolic sense. However, thanks to the analysis of the pollen genome and the profile expression of particular genes, it has been possible to have a greater insight into the unusual transcriptional dynamics of pollen. It was concluded that while ripe pollen is relatively resistant to air pollution in dry air, an increase in humidity triggers a cascade of reactions resulting in activation of highly dynamic metabolic processes, as well as a rapid increase in the process of gene transcription, synthesis of new compounds and conformational modifications of the existing proteins. A classification of transcripts carried out with bioinformatical analysis (DNA-ChipAnalyzer, GeneChip ® Array Genome) has made it possible to identify more than 51,000 compounds, many of which have been classified as PR proteins. In various stages of pollen life, i.e. starting with its creation, then development and germination of pollen tube, there are many changes in gene profiles and expression which take place and which depend on external factors and the toxins present [91]. It can be concluded that plant pollen should be treated as an aerobiological index for atmospheric pollution.

3. Allergens of fungal spores and pollution

Fungal spores are important elements of bioaerosols. Continual climate warming, demonstrated by an increase in temperature, precipitation and floods, causes increased exposure to fungal spores, both in the internal and external environment. Fungal spores have a many-faceted influence; they can induce mycosis, and the proteolytic enzymes and toxic products of metabolism known as called mycotoxins can damage epithelial protective barriers and have pro-inflammatory cancerogenous properties. β-glucans of fungal spores have adjuvant effects which indirectly favour the development of allergies by activating Th2 lymphocytes [67].

Exposure to fungal spores and plant pollen is measured by the volumetric method: measuring the number of spores per $1m^3$. Another method involves measuring the level of fungal products e.g. 1 – 3 β-glucans, which can demonstrate pro-inflammatory properties, both in atopic and non-atopic patients. According to Becker et al [12] the incidence of allergies to fungal spores (unlike to pollen allergens) can be underestimated due to lack of proper standardization of commercial extracts of some fungi such as Aspergillus or Penicillium. The spores of widely known fungi such as Cladosporium, Penicillium, Aspergillus, Helminthosporium and many other species are considered pollutants and infectious factors, as well as important risk factors for the development and aggravation of allergies and asthma. Studies carried out in the north of Poland indicated a minor influence of air pollution on the number of Alternaria and Cladosporium spores in $1m^3$ [51]. Some other studies have shown that pollutants can cause an increase in virulence of fungal spores. It has been confirmed that pollution induces the increase in the synthesis of melatonin in the conidia of Aspergillus fumigatus fungi, which leads to aggregation and makes them more resistant to lysis after being phagocitized by macrophages. The melatonin of mould spores can also contribute to the increase in virulence by not inhibiting apoptosis in macrophages which phagocitose the conidia of fungal spores. It also increases their resistance to ultraviolet rays [139].

Research on Aspergillus fumigatus showed demonstrated the possibility of activating a dozen or so mechanisms of cell signalling, mainly mitogen-activated protein kinase (MAPK). It was concluded that, like in plants, environmental factors activated the genes responsible for the synthesis of PR proteins, enzyme modification and synthesis of pigment which intensified the virulence of fungal spores, induced inflammatory processes and killed experimental animals [1]. Studies carried out in polluted areas in Asia indicated that increases in air temperature, wind speed, rainfall and pollutants (e.g. solid particles – aerodynamic diameter ≤10 μm (PM10) and carbohydrates) increased the number of fungal spores in the atmosphere [24]. People who suffer from low resistance to fungal spores can develop serious infections such as Aspergillosis. It was stated that the exposure of the respiratory system to Aspergillus fumigatus spores can stimulate the production of IFN-γ by specific Th1 (CD4+) lymphocytes or Th2 (CD4+) cytokines, thus favouring allergies, depending on the phenotype of experimental animals [117,84].

The studies showed that the conidia of Cladosporium spp., Stachybotrys spp. and Aspergillus niger are considerably resistant to the activity of pollution, also with ozone [80]. Only high levels of ozone (which increase in cities when the temperature is >30 °C and humidity is low ≥40%) can inhibit the development of fungal spores e.g. Cladosporium spp [39,141].

3.1 Fungal spores can be considered serious air pollutants

It was concluded that the exposure to Aspergillus fumigatus and allergens of house-dust mites favours the development of asthma and the allergic inflammation of the respiratory tract. It thus enhances the activation of congenital and acquired immunological response. Fukahori et al. [47] showed that a strong reaction of dendritic cells to a simultaneous exposure to fungal spores and Der f1 might result from the activation of TLR2 receptors by the β-D glucan of the fungal spores, which in turn, leads to a synergetic activation of dendritic cells by mite allergens. Stimulation or blocking of TLR receptors as well as increases in IL-10 and IL-23 can change the balance of Th1/Th2/Th17/Treg and their influence on the course of allergic reactions. Recently Liu et al [93] presented the results of the exposure of Th2 lymphocytes in mice sensitized to Aspergillus fumigatus allergens and DEP particles. During in vitro studies they noticed epigenetic changes which involved hypermethylation of IL4 and IFNγ gene promoters, which led to polarization of Th lymphocytes toward Th2. Methylation of both the gene promoters was correlated with an increase in IgE level, which might imply that proteins of fungal spores, like air pollutants, can play a role in the pathogenesis of asthma and allergy through epigenetic mechanisms.

4. The environmental pollution and the oxidative stress

4.1 Pathways of toxin penetration.

Particles of dust and toxins are mostly absorbed by the lungs but also by the skin and alimentary tract and even by the placenta, which has a negative influence on a foetus.

4.2 Exposure to toxins of respiratory epithelium.

Human lungs constitute an area of 100 -140 m^2 , they are strongly vascularized and are constantly exposed to infectious factors and environmental pollution. Healthy lungs absorb air at the rate of 14 breaths per minute, which equals about 10 litres per minute and about 10,000 – 15,000 litres per day. Inhaled air, before it gets to the lowest located parts of lungs, is partly filtered and climatized. The mucous membrane in the nose prevents the largest polluted particles from getting inside the body. In order to cleanse themselves, lungs use their cilia, which expel polluted mucosa at 16 beats per second. The respiratory epithelium is a protective barrier which actively protects the body and, if needed, activates the cells of the immunological system. Proteins such as filaggrin, loricrin and involucrin play an important role in reinforcing the epithelium against attack. They make it stronger and more resistant to invasion of microorganisms, allergens and chemical compounds. It was concluded that both epigenetic changes in genes caused by environmental pollution and cytokines of Th2 lymphocytes e.g. IL 4, IL13, TNF-α inhibit the production of filaggrin and other proteins, which might favour immunization, development of atopic dermatitis, allergy and asthma due to increased epithelial permeability [76,102]. In sensitive people, air pollutants, including allergens, can be bound by proteins and lipids of the epithelial cell membrane and thanks to them, can penetrate the inside of the cell.

Epithelial cells can initiate various protective initiatives against air pollution. One example might be the activation of numerous genes (300 – 550). The activation usually leads to the growth of the synthesis of pro-inflammatory cytokines and activation of cells which phagocytose many fractions of dust, allergens and bacteria [101,10]. Studies show that exposure to pollution induces cell stress. This, in turn, activates very complicated mechanisms adjusted to particular situations through increased expression of particular transcriptional factors, the synthesis of protein (e.g. SKN-1/Nrf) and the activity of proteasome in utilizing unneeded particles. These processes facilitate the maintenance of homeostasis of cells and protects them against activation of neoplastic processes and autoaggression [92].

In high levels of pollution and environmental toxins, these natural mechanisms are insufficient. It has been confirmed that the amount of PM particle fractions which penetrate the lungs depends not only on the size of the inhaled particles but also on the ability of the organism to cleanse and detoxify itself, which is genetically determined. However, in people who are exposed to nicotine smoke for a long time, neither detoxifying enzymes nor antioxidative protective mechanisms function properly, which favours bronchial hyperreactivity, asthma and lung-remodelling process.

Figure 4 shows the relative sizes of inhaled particle, patterns of their airway deposition and main health effects of air pollution.

4.3 The environmental pollution and the oxidative stress.

Each organism is constantly exposed to both endogenic and exogenic oxidizing compounds. Endogenous free radicals – reactive oxygen species (ROS), such as: O_2- , OH^*, H_2O_2 and nitrogen - reactive nitrogen species (RNS), e.g. NO^*, $ONOO-$ are constant products of chemical reactions associated with respiration and processes within cells such as activated phagocytes and eosinophils active in allergic processes, and are considered the main source of ROS. In physiological conditions, free radicals pose no threat because of natural enzymatic and nonenzymatic protective systems of an anti-oxidative character [99]. Exogenous free radicals are neutralized mainly in the mitochondria of lung epithelial cells. The key antioxidant is glutathione. The synthesis of protective compounds in relation to ROS is stimulated among others by the activation of transcription factors, e.g. Nrf2, which leads to the activation of proteases and antioxidants and modulates inflammation by the inhibition of the NF-κB pathway. The activation of the Nrf2 metabolic pathway is important in the prevention of many inflammatory diseases, neurodegenerative diseases, diabetes, pulmonary fibrosis and neoplasms, additionally, Nrf2 overexpression in neoplastic cells protects the cells against the cytotoxic activity of antineoplastic drugs [126]. The activation of Nrf2 might ameliorate the degree of lung impairment caused by nicotine smoke [130].

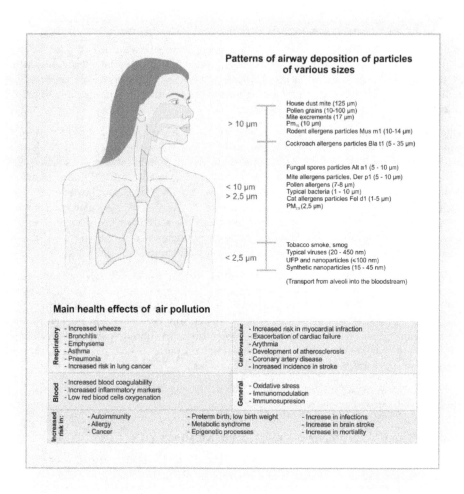

Fig. 4. Relative sizes of inhaled particle, patterns of their airway deposition and main health effects of air pollution. PM - particulate matter, UFP-ultrafine particles (particles <100 nm). Adopted from: Chang C. The immune effects of naturally occurring and synthetic nanoparticles. Journal of Autoimmunity. 2010;34:234-246.

4.4 The oxidative stress and the cascade of inflammatory effects induced by polluted air

The hierarchical model of oxidative stress presented in figure 2 shows the negative effects of the exposure to environmental pollution.

Level 0: Preventing the production of free radicals.

A lot of enzymes prevent the development of chain reactions leading to appearance of free radicals, e.g. glutathione peroxidase.

Level 1: Antioxidative cell protection.

The first stage of protection includes binding free radicals and breaking the cascade reaction. Here antioxidative enzymes and metalloproteins, i.e. proteins which bind ions containing copper and iron, are active. The second stage of protection includes ROS "scavengers": flavonoids, vitamin A, vitamin C, bilirubin, uric acid and glutathione. Relatively mild oxidative stress stimulates cell structures mainly through Nrf2 nuclear transcription factor for transmission signals for the expression of proper genes, production and activation of a numerous group of antioxidants.

Level 2: Inflammation.

If protective mechanisms turn out to be insufficient in neutralizing the excessive number of free radicals, it is difficult to maintain homeostasis. Moreover, an increase in the cascade expression of many pro-inflammatory genes for cytokines, chemokines, adhesive molecules and the heat-shock proteins active in inflammatory processes [71]. The activity of NF-κB transcription factor inhibits Nrf2, which decreases the expression and gene transcription of antioxidants essential in detoxification. The disturbance of the proper function of Nrf2 genes can also impair antigen presentation in the mechanisms of major histocompability complex class II, which makes the body more susceptible to both allergic and infectious diseases [144]. It has also been confirmed that the exposure of a mouse with Nrf2$^{-/-}$ to DEP particles, even to small doses , leads to an increase in bronchial hyperreactivity and oesinophilic inflammation of the respiratory system [87].

Level 3: Damage to mitochondria, cell apoptosis.

The third level of oxidative stress which is impossible to neutralize by protective mechanisms breaks the protective system. Great stress induces changes to Ca2+ ion concentrations, permeability of mitochondrial membrane and disturbances of mitochondrial functions. It contributes to energetic insufficiency of epithelial cells and the release of factors leading to apoptosis. Such processes are induced especially by PM particles and ultra small dust particles (UPF<0.1 µm) collected in urbanized city areas as they are characterized by a higher potential for ROS induction in cells than PM_{10} and $PM_{2.5}$ particles [88]. It is worth mentioning that apoptosis protects the body against the development of neoplasms, especially in the polluted environment.

4.5 Reactive oxygen species (ROS) and bronchial asthma

Excessive exposure to environmental pollution leads to the development of unnecessary ROS and RNS species. They attack cell membranes, disturb their electric potential and proper function of ion pumps. Oxidized cholesterol favours the development of atherosclerotic plaque. Overactive pro-oxidative processes might induce imbalances and induce the oxidizing process. The synergy of endogenous and exogenous anti-oxidants, which might neutralize free radicals, plays a key role in the protection against the development of inflammation of the respiratory passage and aggravation of asthma. The activation of the Nrf2 pathway is to a great extent, proof of adaptation to the natural environment and factors. The anti-oxidative activity of Nrf2 is reduced in patients with severe asthma [104]. Children with severe asthma demonstrated posttranslational modifications of Nrf2 and impaired protection against free radicals. Thus, it was stated that "redox status" and the activity of Nrf2 can be invaluable in the assessment of asthma severity [43].

5. The role of aryl hydrocarbon receptor (AHR) in immunotoxicity

AhRs are considered the most important receptors in the detoxification and discharge of toxins and are represented by a few proteins, including Hsp90 (heat-shock proteins). They are present in the cell cytosol and naturally activated by bilirubin, arachidonic acid, eicosanoids such as prostaglandins, leukotriens, metabolites and some natural compounds present in food (tryptophan derivatives, diet carotenoids). It was discovered that AhRs are activated by over 400 exogenous compounds and that in turn, the activation of an AhR triggers the activation of cell transcription factors.

The TCDD dioxin is one of the strongest AhR ligands as it can securely bind these receptors. Unfortunately, toxins accumulate in the body and, as the half-life for dioxins is between 7 and 10 years, long-term AhR activation might be favoured with a consequential activation of genes encoding the enzymes metabolizing xenobiotics, which induces a chronic suppression of the immune system. Therefore TCDD is regarded as one of the most toxic and carcinogenic environmental toxins [56]. The receptors, having bound to ligands, move to the nucleus, where they activate gene expression , especially those responsible for the synthesis of enzymes involved in the metabolism of toxins, such as superoxide dismutase, glutathione peroxidase and aromatic hydrocarbons hydroxylase (from cytochrome P450 group).

Binding an AhR to a toxin leads first of all, to the development of toxic metabolites. Secondly, it induces changes to gene transcription. Researchers observed stimulation of the transcription of cytokine IL-17, which leads to the recruitment and differentiation of neutrophils, and IL-22, which is essential in the synthesis of defensins necessary in antibacterial immunity. An excess of IL-17 can disturb the balance of Th1/Th2 as well as the balance of Th17/Treg and, consequently, contribute to intensive development of inflammation, autoimmune and neoplastic diseases [150]. Exposure to TCDD increases risk of the development of granulocytic leukemia by up to 400%. The onset of the disease might happen even 15 years later [14]. An excessive amount of toxins might bind to DNA, RNA as well as cell proteins and induce a number of pathologies [54,68,143].

6. Dysfunctions of immune system in the polluted environment

Environmental pollution might cause immunomodulation and the immunosuppression of immune response [128], demonstrating insufficient anti-infectious and anticancer immunity [57]; they may also concern hypersensitivity, including allergy [15] or autoaggression [42]. Research findings indicate that even short-term exposure to toxins in the air can lead to an increase in the incidence of disease and death. It has been estimated that in Canada, air pollution contributed to about 8% of deaths in 2004 and as many as 10% in 2008 [3] The reaction to chemical agents present in the surroundings varies based on the individual.

Genetic factors, age, sex, physical condition, medicines taken, activity of endogenic detoxifying enzymes, past diseases and various factors connected with lifestyle, dietary habits and other risks, such a place of living, exposure to nicotine smoke: they all determine different reactions to environmental chemical agents. Allergy and atopic asthma are also effects of interactions between genetic susceptibility and the polluted environment, which has a great influence on the development of atopic phenotype, especially during rapid development in the prenatal and postnatal period [58]. Intensive exposure to allergens might have an impact on the development of tolerance. Yet, coexistence of toxins usually causes adjuvant activity. Early exposure to toxic substances, e.g. dioxins from nicotine smoke, PM particles, DEP, ozone might stimulate and preserve the Th2 atopic phenotype dominant in pregnancy. The process can be accompanied by suppression of Th1 cell proliferation. It might lead to the increased risk of the development of atopic asthma in children [96].

6.1 The examples of pollutants influence on the immunological response

6.1.1 Coal compounds

Coal destroys ecosystems and pollutes the environment when it is transported, stored and burnt. Also storing waste has negative implications on the environment. The process of burning coal is considered the main cause of global warming as it is responsible for the 30 % growth of CO_2 and methane (CH_4) in the atmosphere compared to the pre-industrial age [116]. Toxic substances in waste and ash (e.g. chromium, cadmium, mercury, arsenic) can contaminate potable water and food, and volatile fractions of ash can emit radiation 100,000 times stronger which nuclear power stations producing the same amount of energy.

Health status is the ability of the body to adapt to changeable surrounding conditions. However, stress induced by excessive exposure to exhaust fumes and environmental toxins is directly connected with the development of chronic inflammation of lower air passages and even leads to the increase in death rate. A memorable smog which happened in Belgium in 1930 [106] and in London in 1952, and a high level of SO_2, which lasted for 5 days in the smog, killed thousands of inhabitants of the city. At that time, the energy source was coal. In Silesia, Poland, where coal was mined on a relatively high scale in Europe, the mortality rate for young men (until 1989) was about 40% higher than the average national mortality rate . Also in Germany between 1995 and 1998, researchers observed a higher mortality rate which was closely connected with air pollution [111]. Sang et al. [121] confirmed that an increased risk of hospitalization and death resulting from SO_2 exposure might be connected with its influence on excessive activation of Cyclooxygenase-2 (COX-2) and prostaglandin PGE2, which leads, among other things, to the development of inflammatory state and even neurotoxic activity.

Carbon monoxide (CO) appears as the result of exhaust fume emission, fires and volcanic eruptions. If coal is burned properly, the emitted fumes contain about 1% CO. a safe level for a human. In the case of insufficient oxygenation, fumes can contain as much as 30% CO. The compound formed by binding hemoglobin and CO is called carboxyhaemoglobin, and the process not only occurs 200 times more readily than in the case of oxygen, but the product is more durable. A thirty-minute exposure to 0.1% – 0.2 % CO leads to death.

6.1.2 Nitrogen compounds

In the 20th century, the amount of nitrogen introduced to the cycle in the biosphere more than doubled and the process is still continuing. Nowadays, apart from pollutants characteristic of "the age of coal and steel" (mainly SO_2 and PM) many other products of burning crude oil and its derivatives used in industry, as well as road and air transport exist. They have been emitted for more than 200 years and their level is constantly increasing. The transport sector is the second greatest source of CO_2 emission in the EU; in the years 1990 – 2009, its emission increased by 20%. However, these fuels are not burnt completely and apart from CO_2, ever more toxic volatile and reactive nitrogen species (RNS) known as NOx, O_3, SVOC are being released into the atmosphere. It is estimated that transport is responsible for about 75% of the nitrogen oxide pollution . Apart from emitting pollutants, vehicles raise dust into the air , which might lead to greater bioavailability of allergens of plant pollen and fungal spores, as well as particles of latex from car tyres. As a result, there might be an increase in allergies to plant pollen, fungal spores as well as latex [116].

The natural level of nitrogen oxide ranges from 0.1 to 9.4 $\mu g/m^3$. The WHO defines a 1-h limit value of 200μg NO_2/m^3 air, equivalent to 0.1ppm NO_2. However, recent studies showed that thirty-minute exposure of nasal epithelial cells in tissue culture to 0.1ppm NO_2 resulted in cytotoxicity and genotoxicity [78]. NO_2 levels ranging from 100 ppb to 400 ppb may impair proper functioning of lungs and the level above 5ppm might lead to damage. [65] In many industrial cities, the NO_2 level is about 2 ppm [73]. It was concluded that NO_2 level in the air outside depends on population density and intensity of transport. [17]

Due to their poor water solubility, the site of toxicity of nitric oxides is the upper respiratory tract. Studies carried out on healthy volunteers demonstrated that a two-hour exposure to transport pollutants analyzed at an underground station in the centre of Stockholm (in peak hours) led to the increase in the expression of markers of Treg cells and activation of fibrinogen in the blood [119] Jeng et al [69] confirmed that a five-day constant exposure to PM 10 particles resulted in an increase in fibrinogen level and IL6 level (i.e. cytokines which stimulate B-lymphocyte differentiation to plasma cells and synthesis of antibodies). But for people with asthma, even a short exposure to transport pollutants might turn out to be hazardous. Sixty patients with asthma were naturally exposed to city pollution for two hours by walking along Oxford Street in London. The studies on short-term exposure to city exhaust fumes demonstrated that, despite the lack of subjective symptoms, forced vital capacity (FVC) and one second forced expiratory volume (FEV1) values were considerably reduced in comparison with the values gathered from the control group. Apart from those different values, an increase in marker level in sputum (such as mycloperoxidase marker - MPO) and a decrease in pH in respiratory passages were also observed [103].

6.1.3 Particulate matters (PM)

Particulate matters (PM) are emitted directly to the atmosphere (primary PM) or produced from gas precursors, e.g. SO_2, NO_x, NH_3, VOCs (secondary PM). PM is represented by solid and liquid particles which can be of organic or inorganic character. It can be inhaled and accumulated in the lungs. Exhaust fume emission, burning fuels, refining processes and other kinds of environmental contamination contribute to the production of PM, which is usually classified on the basis of particle size. The following fractions are monitored most frequently: PM_{10} (< 10μm), $PM_{2.5}$ (<2.5 μm), $PM_{0.5}$ (<500nm), $PM_{0.1}$ (<100nm). PM_{10} are usually spherical, aerodynamic particles whose diameter is about 10μm. The toxicity of PM depends mostly on the chemical composition of the particles. Exceeded level of PM is considered an air quality index in EU countries. In cities, fractions of particles are mostly <2.5 μm, and over 80% of fractions have a diameter of <0.1 μm. The new trend, however, sounds optimistic; between 1990 and 2008 emission of PM particles decreased in European countries on average by 21%, but slight increase in the fractions of PM 2.5 – 10 μm was also observed [61]. Study findings carried out in Łódź in the centre of Poland in 2009 showed the maximum daily amount of PM_{10} being exceeded 35 times [64]. Research conducted in China and India indicates that the increase in PM_{10} level in city air by 10 μg/m³ contributed to an increase in mortality rate by 0.6% [27].

$PM_{2.5}$ might induce some extra disorders in the cardiovascular system. Very small particles coming from exhaust fumes can enter the bloodstream and reach distant organs such as the heart and kidneys. There is also the possibility of micro-injuries of endothelial cells appearing in blood vessels which might affect the activation of coagulation factors and impair autonomic regulation of heart rate, which in turn, may cause sudden death because of arrhythmia [44].

The main cell lines involved in immunological response are epithelial cells, dendritic cells, Th lymphocytes, macrophages, neutrophils, mast cells, basophils and eosinophils. They protect the body's well-being but in a polluted environment, these cells can also get involved in the cascade of events leading to inflammation and various pathological forms. Many studies indicated that exposure to PM_{10} particles exacerbates symptoms in patients with asthma, contributes to increased administration of anti-asthmatic medications and frequent visits to emergency medical services [7]. Particles appearing in city and industrial areas are especially dangerous. Studies on air aerosol gathered in Bejing, with a population of twelve million , indicated that over 99% of PM particles had a diameter smaller than 1μm and the endotoxin level can reach a value of about 1250 pg/mg of aerosol. High immunomodulating, toxic and even mutagenic potential might be caused by the presence of many toxins and their easy deposition in lungs owing to very low weight [142]. It was concluded that they can inhibit the synthesis of IFNγ in human leukocytes of peripheral blood. As a result, the balance of the Th1/Th2 leukocyte population is disturbed. The dominant leukocytes are Th2 leukocytes which are responsible for the development of allergy. Simultaneously, anti-infectious and anti-neoplastic immunity gets reduced. Dockery et al [35] confirmed the relationship between $PM_{2.5}$ in the air and death rate. Cohort studies conducted in 6 cities in the US showed that $PM_{2.5}$ increase by about 10g/m³ in a year entailed an increase in death rate ranging from 10.9% to 20.8% [120]. Rezentiti et al [114] concluded that transferring children living in cities to unpolluted rural areas contributes to quick and considerable improvement of aspects of their health state such as well-being, peak expiratory flow (PEF) and monitored inflammation indicators (the decrease in the number of oesinophils in nasal lavage and the level of nitrogen oxide in exhaled air). Studies performed in Bejing indicate that extremely small particles stimulate pulmonary oesinophilia and the activation of Th2 response, which might be connected with the presence of β-glucan of fungal spores in inhaled urban particulate matter (UPM) [53].

6.1.4 Nanoparticles (ultrafine particles UFP)

Nanoparticles (ultrafine particles UFP). They belong to a class of the smallest PM particles: <100 nm. Two types of nanoparticles are recognised: natural nanoparticles emitted in the process of incomplete burning by diesel engines, and synthetic ones (produced deliberately) which have medical applications (e.g. in antibacterial nanocoating), electronic engineering and other industries. Many scientists stress their exceptional ability to penetrate the deepest parts of lungs and the inside of the cell. Research performed with the use of an electron microscope confirms that nanoparticles can damage even the internal structures of mitochondria [89]. They can easily get to the cardiovascular system and cause arrhythmia by changing cardiomyocyte contractility [145]. It was concluded that they can have some pro-inflammatory properties, raise blood pressure and favour atherosclerosis [8]. During exposure to UFP, mononuclear cells are chemotactically activated. The exposure also induces the expression of pro-inflammatory genes in macrophages and endothelial cells and stimulates the synthesis of pro-inflammatory cytokines, e.g. IL1B, TNF, IL4 [48]. The particles act through congenital and acquired immunity mechanisms and influence the development of Th2-type allergic inflammation of the upper respiratory tract [75]. As they are relatively large and light, they absorb many organic compounds, making them even more toxic. Li et al [90] stated that chemical composition of UFP, emitted by high-pressure diesel engines in trucks, was different for different working cycles of the engines. UFP particles emitted by stationary trucks directly induced oxidative stress in human aorta endothelial cells, and UFP particles emitted while the truck was moving contained four times more metal ions (iron, chromium and nickel) and more intensively activated inflammatory processes such as the increase in the gene expression for IL-8 chemokines and monocyte chemoattractant protein -1 (MCP-1) as well as adhesive molecules.

6.1.5 Diesel Exhaust Particles (DEP) – Toxin transporters

Diesel Exhaust Particles (DEP) – toxin transporters. Exhaust fumes which appear in the process of combustion of fuel oil tend to aggregate into separate, spherical particles which have a diameter of 0.1-0.5μm. The core of the DEPs are tiny specks of soot which are mixed in with hundreds of organic and inorganic substances. The influence of DEP particles on the immune system depends on the level of exposure, the particle diameter, their chemical composition, and their reactions with other chemical compounds, both organic and inorganic, including allergens and bacterial endotoxins. Experimental and epidemiological research shows that DEPs in city air have a more toxic action than those in a rural area. What's more, their effects can be reversible or irreversible, as manifested by various pathologies such as damage to epithelial cells, pulmonary emphysema, allergic asthma, cardiological disorders and even acute renal failure [107]. Studies have addressed the many mechanisms responsible for adverse immunological DEP-induced response and have demonstrated that that epithelial cells in the respiratory system are the first to detect pollutants and react to them, as DEPs can diffuse through the cell membrane and bind to receptors in the cytoplasm. One of such receptors is the aryl hydrocarbon receptor (AhR). Internalization of DEPs to the inside of the cells activates numerous signalling pathways which induce, inter alia, the synthesis of granulocyte-macrophage colony stimulating factor (GM-CSF) in the respiratory tract. This is essential in asthma pathogenesis as inducing the differentiation process of hematopoietic cells into precursors of granulocyte, macrophage and eosinophil lines favours proinflammatory changes in lungs. Having been exposed to

DEPs , the epithelium releases a number of mediators which have chemotactic properties and influence dendritic cells. These, in turn, activate naive T lymphocytes, which causes a proliferation of their clones in various effector lines. Of the examined air pollutants, DEPs from city areas are considered most allergenic [148]. A body of research shows that exposure of dendritic cells to PM particles stimulates Th to produce IL13 and IL1 and inhibits IFNγ. DEP extracts induce oxidative stress in bone marrow dendritic cells, which causes disturbances in the production of IL-12 cytokine, which plays a key role in the process of Th1 lymphocyte proliferation. Exposure in the early stage of life may be critical for the stability of the Th2 phenotype. As research demonstrates, DEP-induced oxidative stress contributes to the activation of Th lymphocytes, induction of Th2 cytokine and sIgE synthesis as well as the proliferation of eosinophils, release of inflammatory mediators, and goblet cell hyperplasia. Nasal administration of DEP solution performed in healthy non-smokers at a dose of 0.3 mg (daily corresponding dose in Los Angeles) induced an increase in IgE level in nasal lavage [34].

Many experiments have confirmed that DEPs modify the immunological response in relation to allergens and induce immunological reactions in the respiratory tract, even in response to low levels and short exposures [94]. They also stimulate increased expression of cell adhesion (ICAM1 and VICAM1), which might lead to thrombosis, acute ischemic stroke and an increase in cytokine levels, i.e. IL6 and IL10 as well as histamine and also the increase in the number of neutrophils and platelets [129]. In patients with asthma, DEP exposure leads to the development of metacholine-induced bronchial hyperreactivity. It was confirmed that DEPs can weaken antibacterial immune functions. In animal experiments DEPs proved to inhibit the expression of major histocompability complex class II molecules which are on antigen presenting cells and inhibit the ability of bacterial antigen presentation [32]. They also inhibit the activation of T lymphocytes and the T-dependent immune response to the dangerous zoogenous bacteria Listeria monocytogenes. They also modulate IL10 synthesis, which acts as an inhibitory factor for immune response. In such conditions, bacteria living in phagocytes can easily proliferate [149].

6.1.6 Policyclic aromatic hydrocarbons (PAHs)

Policyclic aromatic hydrocarbons (PAHs) are benzene derivatives such as xylene, toluene, pyrene and benzpyrene. They penetrate cell membranes, accumulate in the lipids of various tissues, mainly in the liver and bone marrow. According to research , cigarette smoke is an important source of PAHs. PAH metabolites were traced in pregnant women living in city areas in Canada [108]. They had strong immunosuppressive properties which, in experimental animals, were represented by pancytopaenia, inhibition of cell immunity and also humoral, i.e. IgA, IgE deficiency, in response to thymus-dependent and thymus-independent antigens, deficiency of proteins of the complement system as well as the impairment of phagocytosis connected with blocking Fc receptors. Moreover, it was confirmed that PAHs influence human B lymphocytes and stimulate the effects of class switching of antibodies, which enhances IgE synthesis [132]. Metabolites of these compounds are also hazardous as they are characterized by genotoxic properties [1,49].

6.1.7 Dioxins

Dioxins – a general name of more than 200 chemical compounds from the group of chlorinated hydrocarbons. They are built of two benzene rings connected with two oxygen bridges. They are produced in the burning process of plastic products, the production of herbicides and disinfectants and while smoking cigarettes. In the air and soil they are not diluted with rainwater but they accumulate in the body in lipids. They are known as "the most toxic chemical compounds ever produced by man". For example, dioxin TCDD – tetrachlorodibenzodioxin is known for its strong carcinogenic properties. Experiments on animals showed that dioxins demonstrate immunotoxic properties. The effects of TCDD exposure merit research as these compounds, even in low concentrations, demonstrate a wide spectrum of activities and are hence known as "environmental hormones". Dioxins have been proven to decrease the expression of major histocompability complex class II and exposure to larger doses causes atrophy and suppression of the thymus – a place where T lymphocytes mature [134]. When adsorbed,dioxins on PM particles reach the lungs they are phagocytized by pneumocytes – epithelial cells of the respiratory system. From there they are transported by blood to the liver where they bind to an AhR receptor. The exposure of animals even to minimal doses of TCDD (0.01lg/kg of body weight) caused a stable binding of the compound to the AhR receptor, which inhibited the proliferation of bone marrow stem cells and activated the immunosuppression [13]. They can also bind to estrogen receptor (ER), which might disturb the activity of estrogens and other hormones, and also induces degradation of ER receptors [52]. These toxins, depending on the size of the dose, have an impact on CD4+ lymphocytes and stimulate their differentiation to T-helper as well as Th1, Th2 and Th17 effector lymphocytes. They also promote the development of suppressive Treg lymphocytes.

Although cell proliferation is considered an irreversible process, it has recently been discovered that dioxins such as TCDD can inhibit the differentiation of B cells into plasmatic cells by changing the expression of transcription factors. Large doses of these dioxins can contribute to the change in programming completely differentiated plasmatic cells back to B lymphocytes. The effects of the change in programming mature B lymphocytes to the cells which have progenitor phenotype [16]. It was stated that they might disturb proper functioning of steroids. They also play a role in the modulation and synthesis of many important cytokines, e. g. TNFα, IL1, IL6, TGFα, TGFβ and IL4 and the exposure to dioxins in a pre-natal period induces hazardous immunosuppression [85,131,23]. Dioxins were proved to disregulate the expression of many genes, among others, in epithelial cells of the respiratory passage, which contributes to the occurrence of many genotoxic symptoms, such as production of unusual proteins, including ubiquitins, and these in turn, disturb transcription processes, favour autoimmunization and the development of neoplasms in fetuses of experimental animals [55,70,140].

6.1.8 Ozone (O₃)

Ozone (O_3) - is a form of oxygen with three atoms in a molecule. The ozone layer which spreads at the height 10 – 50 km from the ground positively contributes to thermal balance of the Earth and absorbs 99% of UV radiation, which has fateful implications for all living organisms, although the same wavelength is responsible for vitamin D production. O_3, unlike other toxins, is a secondary pollutant, since it is produced from the hydrocarbons and nitrogen oxides of which air is composed with the use of natural sunlight. The half-life for tropospheric ozone is around 22 days [63]. Ozone is a strong oxidizing, irritating and bactericidal agent. The reaction of ozone with water molecules leads to its hydrolysis and the production of free radicals. Reactions of O_3 with air pollutants change the chemical composition of smog. They enhance its irritating properties for the eyes, mucous membrane of the respiratory passage.

Being a strong oxidant, ozone can react with many biomolecules inside and outside a cell. It disturbs the metabolism and photosynthesis of plants. It has a negative influence on the respiratory and alimentary systems and also on the skin. It was stated that ozone might have caused more than 20, 000 premature deaths and about 200 million episodes of respiratory failure within only one year [136]. Many authors also stress that O_3 and symptoms of allergy, asthma and chronic obstructive pulmonary disease (COPD) are closely related [33,46,77]. The early exposure to ozone might favour the Th2 atopic phenotype, which leads to a more probable risk of the development of allergy and asthma in children. Kopp et al [79] conducted studies in which they monitored allergy symptoms in 170 children. They concluded that the symptoms were aggravated as the ozone level rose, however, after some time, the symptoms became more stable despite the ozone level remaining relatively high. The researchers showed that frequent inhalations with 0.5 ppm of ozone and house dust mite (HDM) carried out for six months in young atopic monkeys induced the amplification of immune response , the increase in IgE levels, oesinophilia, changes in the respiratory tract, an increase in bronchial hyperreactivity and histamine level in serum. Inhalations performed exclusively with ozone lead to the increase in histamine level in serum, probably on the IgE-independent way [122]. According to Peden et al [109] ozone induces pro-inflammatory effects with accompanying mechanisms not examined well yet (in comparison to other air pollutants) in which monocytes and macrophages take part. Ozone was also confirmed to increase P substances in the respiratory tract [82].

7. Epigenetic changes and environmental pollution

Epigenetic modifications are considered a bridge connecting the genotype and phenotype of living organisms. Air pollutants can contribute to many epigenetic changes, by changes to gene expression, e.g. by DNA methylation and changes to histone structure. They also might facilitate the transcription activation of promoters and modification of RNA segments or make these processes difficult to occur. The dynamics of epigenetic modifications are varied [6]. It is believed that some genes may become activated only if they are affected by environmental exposure, for example to nicotine smoke, as well as by PAHs which induce chromosome aberrations and oncogenic mutations. [110]. It can be concluded that epigenetic changes induced by the polluted environment play a key role in the development of asthma and allergic diseases. They can modify transcription of genes connected with immune reactions, induce pro-inflammatory response and simultaneously cause remission or aggravation of the disease and even control the effectiveness of pharmacological treatment [11,83]. These changes are usually reversible but they can also be hereditary. Research is being performed on epigenetic biomarkers essential in the early stage of life and their role in the development of asthma and allergy in later life, as well as the relationship between the findings of clinical observations made during exposure to pollutants and gene expression for many important factors, e.g. Treg cells.

8. Conclusions

A polluted atmosphere is only one of many examples of destruction of the natural environment. There is sizeable body of evidence confirming the considerable anthropogenic influence that exists on air composition and even the climate on the Earth. These human activities influence all forms of life, including the life of man himself. Our immune system stays alert and strives to minimize the risk of exposure to hazardous level of air pollutants which might posses synergistic and antagonistic properties. The degree of sensitivity to air pollutants is different for each individual. It depends on age, health condition and genetic factors. It is therefore difficult to determine the maximum levels of environmental toxins. We are ever more aware of the source of the problems of environmental pollution. Nevertheless, it is difficult to solve them. Reasonable pro-health education seems to contribute considerably to the improvement of air quality and the condition of the natural environment. The mentioned problems should be solved on a world scale.

9. Acknowledgments

We thank Dorota Wawrzyniak for translation of manuscript and Agnieszka Wojciechowska for all pictures. We are also grateful for the time and efforts of Edward Lowczowski in reviewing this manuscript.

10. References

[1] Abad A, Fernández-Molina JV, Bikandi J, et al. What makes Aspergillus fumigatus a successful pathogen? Genes and molecules involved in invasive aspergillosis. Rev Iberoam Micol. 2010 27:155-82.

[2] Abbas I, Garcon G, Saint-Georges F, et al. Polyciclic aromatic hydrocarbons within air borne particulate matter (PM (2.5) produced DNA bulky stable adducts in a human lung cell coulture model.

[3] Abelsohn A, Stieb DM. Health effects of outdoor air pollution: Approach to counseling patients using the Air Quality Health Index. Can Fam Physician. 2011 57:881-7.

[4] Ahire KC, Kapadnis BP, Kulkarni GJ et al. Biodegradation of tributyl phosphate by novel bacteria isolated from enrichment cultures. Biodegradation. 2011. DOI 10.1007/s10532-011-9496-7.

[5] Allard-Coutu A, Martin J, Shalaby K. Inhaled pollen-induced airways disease depends on oxidative stress but not Toll-like receptor 4. American Thoracic Society International Conference.2011; abstract. https://mysts2011.zerista.com.

[6] Alvarez R, Altucci L, Gronemeyer H, de Lera AR. Epigenetic multiple modulators. Curr Top Med Chem. 2011. PMID:22039877.

[7] Andreson HR, Ponce de Leon A, Bland JM, et al. Air pollution, pollens, and daily admission for asthma in London 1987-92. Thorax 1998;53:842-848.

[8] Araujo JA, Barajas B, Kleinman M, et al. Ambient particulate pollutants in the ultrafine range promote early atherosclerosis and systemic oxidative stress. Circ Res. 2008;102:589-596.

[9] Ariano R, Canonica GW, Passalacqua G. Possible role of climate changes in variations in pollen seasons and allergic sensitizations during 27 years. Annals of Allergy, Asthma & Immunology. 2010; 104: 215-222.

[10] Ather JL, Alcorn JF, Brown et all. Distinct functions of airway epithelial nuclear factor-kappaB activity regulate nitrogen dioxide-induced acute lung injury. Am J Respir Cell Mol Biol. 2010;43:443-51.

[11] Bastonini E, Verdone L, Morrone S, et al. Transcriptional modulation of a human monocytic cell line exposed to PM(10) from an urban area. Environ Res. 2011;111:765-74.

[12] Becker WM, Vogel L, Vieths S. Standardization of allergen extracts for immunotherapy: where do we stand? Curr Opin Allergy Clin Immunol 2006;6:470-5.

[13] Bemis JC, Alejandro NF, Nazarenko DA, et al. TCDD-induced alterations in gene expression profiles of the developing mouse paw do not influence morphological differentiation of this potential target tissue. Toxicol Sci. 2007;95:240-248.

[14] Bertazzi PA, Consonni D, Bachetti S, et al. Health effects of dioxin exposure: a 20-year mortality study. Am J Epidemiol. 2001;153:1031-1044.

[15] Bezemer GF, Bauer SM, Oberdörster G, et al. Activation of pulmonary dendritic cells and Th2-type inflammatory responses on instillation of engineered, environmental diesel emission source or ambient air pollutant particles in vivo. J Innate Immun. 2011;3:150-66.

[16] Bhattacharya S, Conolly RB, Kaminski NE, et al. A bistable switch underlying B-cell differentiation and its disruption by the environmental contaminant 2,3,7,8-Tetrachlorodibenzo-p-dioxin.Toxicol. Sci. 2010;1:86-97.

[17] Blomberg A, Krishna MT, Bocchino V, et al. The inflammatory effects of 2 ppm NO2 on the airways of healthy subjects. Am. J. Respir. Crit. Care Med.1997;156:418–424.

[18] Bömmel H, Min Li-Weber M, Serfling E et al. The environmental pollutant pyrene induces the production of IL-4. The Journal of Allergy and Clinical Immunology. 2000; 105:796-802.

[19] Breiteneder H, Ebner C. Molecular and biochemical classification of plant-derived food allergens. J Allergy Clin Immunol. 2000;106:27-36.

[20] Brondz I, Brondz A. Suppression of immunity by some pesticides, xenobiotics, and industrial chemicals. In vitro model. Journal of Biophysical Chemistry. 2011; 2, doi:10.4236/jbpc.2011.23028.

[21] Burney PGJ, Newson R B, Burrows M S, et al. The effects of allergens in outdoor air on both atopic and nonatopic subjects with airway disease. Allergy 63, 2008:542-546.

[22] Butz AM, Matsui EC, Breysse P, et al. A randomized trial of air cleaners and a health coach to improve indoor air quality for inner-city children with asthma and secondhand smoke exposure. Arch Pediatr Adolesc Med, 2011;165:741-748.

[23] Całkosiński I, Stańda M, Borodulin – Nadzieja L, et all. The Influence of 2,3,7,8−Tetrachlorodibenzo− p−Dioxin on Changes of Parenchymal Organs Structure and Oestradiol and Cholesterol Concentration in Female Rats. Adv Clin Exp Med. 2005;14: 211-215.

[24] Chao HJ, Chan CC, Rao CY, et al. The effects of transported Asian dust on the composition and concentration of ambient fungi in Taiwan. Int J Biometeorol. 2011; PMID:21328007.

[25] Chehregani A, Kouhkan F. Diesel exhaust particles and allergenicity of pollen grains of Lilium martagon. Ecotoxicology and Environmental Safety. 2008;69:568-573.

[26] Chehregani A, Majde A, Mostafa Moin M, et al. Increasing allergy potency of *Zinnia* pollen grains in polluted areas. Ecotoxicology and Environmental Safety. 2004;58:267-272.

[27] Chung KF, Zhang J, Zhong N. Respirology. Outdoor air pollution and respiratory health in Asia. Respirology.2011;16: 1023-1026.

[28] Clot B, Trends in airborne pollen: an overview of 21 years of data in Neuchâtel (Switzerland). Aerobiologia. 2003;19:227-234.

[29] Colliar L, Sturm A, Leaver MJ. Tributyltin is a potent inhibitor of piscine peroxisome proliferator-activated receptor α and β. Comp Biochem Physiol C Toxicol Pharmacol. 2011;153:168-73.

[30] Corsini E, Oukka M, Pieters R, et all. Alterations in regulatory T-cells: Rediscovered pathways in immunotoxicology. J Immunotoxicol. 2011; PMID: 21848365.

[31] Deifl S, Bohle B. Factors influencing the allergenicity and adjuvanticity of allergens. Immunotherapy. 2011;3: 881-893.

[32] Devalia JI, Bayram H, Abdelaziz M, et al. Differences beetween cytokine release from bronchial epitelial cells of asthmaic patients and non-asthmatic subjects : effect of exposure to diesel Exhaust particles. Int Arch Allergy Immunol 1999;118:437-439.

[33] Di Giampaolo L, Quecchia C, Schiavone C, et al. Environmental pollution and asthma. Int J Immunopathol Pharmacol. 2011;24 (1 Suppl):31S-38S.

[34] Diaz-Sanchez D, Dotson AR, Takenaka H, et al. Diesel Exhaust particles induce local IgE production in vivo and alter the pat tern of IgE production in vivo and alter the pat tern of IgE Messenger RNA isoforms. J Clin Invest 1994;94:1417-25.

[35] Dockrey DW, Schwartz J, Spengler JD. Air pollution and daily mortality association with particles and acid aerosols. Environmental Res. 1992;59:362-373.

[36] Dubey PS, Mall LP. Herbicidal pollution. Pollen damage by herbicyde vapours. 1971.

[37] Dyrektywa CAFE 2008/50/WE Parlamentu Europejskiego i Rady z dnia 21 maja 2008r. w sprawie jakości powietrza i czystszego powietrza dla Europy (Dz. Urz. UE L. 152 z 11.06.2008). Dyrektywa CAFE 2004/107/WE Parlamentu Europejskiego i Rady z dnia 15 grudnia 2004 r. w sprawie arsenu, kadmu, rtęci, niklu i wielopierścieniowych węglowodorów aromatycznych w otaczającym powietrzu (Dz. Urz. UE L 23 z 26.01.2005:3).

[38] Eckl-Dorna J, Klein B,Reichenauer T, et al. Exposure of rye (Secale cereale) cultivars to elevated ozone levels increases the allergen content in pollen. The Journal of Allergy and Clinical Immunology. 2010;126: 1315-1317.

[39] Elminir HK. Dependence of urban air pollutants on meteorology. Science of the Total Environment 350. 2005:225-237.

[40] Emara AM, Abo El-Noor MM, Hassan NA, et al. Immunotoxicity and hematotoxicity induced by tetrachloroethylene in egyptian dry cleaning workers. Inhal Toxicol. 2010;22:117-24.

[41] Esplugues A, Ballester F, Estarlich M, et all. Indoor and outdoor concentrations and determinants of NO_2 in a cohort of 1-year-old children in Valencia, Spain. Indoor Air. 2010 20:213-23.

[42] Farhat SC, Silva CA, Orione MA, et all. Air pollution in autoimmune rheumatic diseases: A review. Autoimmun Rev. 2011. doi:10.1016/j.autrev.2011.06.008.

[43] Fitzpatrick AM, Stephenson ST, Hadley GR, et al. Thiol redox disturbances in children with severe asthma are associated with posttranslational modification of the transcription factor nuclear factor (erythroid-derived 2)-like 2. J Allergy Clin Immunol. 2011 ;127:1604-11.

[44] Franchini M, Mannucci PM. Short-term effects of air pollution on cardiovascular diseases: outcomes and mechanisms. Journal of Thrombosis and Haemostasis. 2007;11: 2169–2174.

[45] Frei T, Gassner E. Trends in prevalence of allergic rhinitis and correlation with pollen counts in Switzerland. Int J Biometeorol. 2008;52: 841-847.

[46] Frush S, Li Z, Potts EN, Du W, et al. The role of the extracellular matrix protein mindin in airway response to environmental airways injury. Environ Health Perspect. 2011;119:1403-8.

[47] Fukahori S, Matsuse H, Tsuchida T, et al. Aspergillus fumigatus regulates mite allergen-pulsed dendritic cells in the development of asthma. Clinical & Experimental Allergy. 2010;10:1507-1515.

[48] Ganguly K, Upadhyay S, Irlmler M, et al. Impaired resolution of inflamatory response in the lungs of JF1/Msf mice following carbon nanoparticle instillation. Respr Res 2011;12:94.

[49] Georgiadis P, Kyrtopoulos SA. Molecular epidemiological approaches to the study of the genotoxic effects of urban air pollution. Mutat Res. 1999;428:91-98.

[50] Ghosh D, Chakraborty P, Gupta J. Asthma-related hospital admissions in an Indian megacity: role of ambient aeroallergens and inorganic pollutants. Allergy; 2010; 65:795-796.

[51] Grinn-Gofroń A, Strzelczak A, Wolski T. The relationships between air pollutants, meteorological parameters and concentration of airborne fungal spores. Environmental Pollution. 2011, 159: 602-608.

[52] Harper JW. Chemical biology: A degrading solution to pollution. Nature 2007;446, 499-500.

[53] He M, Ichinose T, Yoshida S, et al. Urban particulate matter in Beijing, China, enhances allergen-induced murine lung eosinophilia. Inhalation Toxicology. 2010; 22: 709–718.

[54] Hemminiki K, Grzybowska E, Chorąży M. DNA adducts In humans environmentally expose to aromatic compounds In an industrial area of Poland. Carcinogenesis 1990,11,1229-1231.

[55] Holladay SD, Mustafa A, Gogal RM. Prenatal TCDD in mice increases adult autoimmunity. Reprod Toxicol. 2011;31:312-318.

[56] Houge C. More dioxin delays. Chemical & Engineering News. 2010;88; 46:30-32.

[57] Hrubá E, Vondráček J, Líbalová H, et al. Gene expression changes in human prostate carcinoma cells exposed to genotoxic and nongenotoxic aryl hydrocarbon receptor ligands. Toxicol Lett. 2011; 10;206:178-88.

[58] http//www.esmo.org/events/lung-2010-iaslc.html.

[59] http: www.cas.org

[60] http://www.airnow.gov.

[61] http://www.eea.europa.eu/about-us/who

[62] http://www.lungusa.org/. Facts about Indoor Air Pollution.

[63] http://www.wiklipedia.org;za Stewenson et al. 2006.

[64] http://www.wios.lodz.pl. Roczna ocena jakości powietrza w województwie łódzkim w 2009 r. Wojewódzki Inspektorat Ochrony Środowiska w Łodzi.

[65] Huss-Marp J, Darsow U, Brockow K, et al. Can Immunoglobulin E-measurement replace challenge tests in allergic rhinoconjunctivits to grass pollen? Clin Exp Allergy. 2011;41:1116-24.

[66] Inoue K, Koike E, Takano H, et al. Effects of diesel exhaust particles on antygen presenting cells and antigen-specific Th immunity in mice. Exp Biol Med (Maywood). 2009;234:200–209.

[67] Inoue Y, Matsuwaki Y, Shin SH, Ponikau JU, Kita H: Nonpathogenic, environmental fungi induce activation and degranulation of human eosinophils. J Immunol 2005, 175:5439-5447.

[68] Izdebska-Szymona K. Immunotoksykologia – nowa gałąź immunologii. Polish Journal of Immunology;18; 1993:223-237.

[69] Jeng HA. Chemical composition of ambient particulate matter and redox activity. Environmental Monitoring and Assessment. 2010;169: 597-606.

[70] Jin KS, Park CM, Lee YW. Identification of differentially expressed genes by 2,3,7,8-tetraclorodibenzeno-p-dioxin in human bronchial epithelial cells. Exp Toxicol. 2011;1. PMID:

[71] Jung EJ, Avliyakulov NK, Boontheung P, Loo JA, Nel AE. Pro-oxidative DEP chemicals induce heat shock proteins and an unfolding protein response in a bronchial epithelial cell line as determined by DIGE analysis. Proteomics 2007;7:3906–3918.

[72] Karl T, Harley P, Emmous L et al. Efficient Atmospheric Cleanising of oxidized organic trace gases by vegetation. Science 2010;330:816-819.

[73] Kattan M, Gergen PJ, P. Eggleston, C.M. Visness and H.E. Mitchell, Health effects of indoor nitrogen dioxide and passive smoking on urban asthmatic children. J. Allergy Clin. Immunol. 2007; 120:618–624.

[74] Kawakami T, Isama K, Matsuoka A. Analysis of phthalic acid diesters, monoester, and other plasticizers in polyvinyl chloride household products in Japan. J Environ Sci Health A Tox Hazard Subst Environ Eng. 2011;46:855-64.

[75] Ken-Ichiro I, Takano H. Facilitating effects of nanoparticles/materials on sensitive immune-related lung disorders. Journal of Nanomaterials; 2011. Article ID 407402,1- 6; doi:10.1155/2011/407402.

[76] Kim BE, Howell MD, Guttman E, et al. TNF-α Downregulates Filaggrin and Loricrin through c-Jun N-terminal Kinase: Role for TNF-α Antagonists to Improve Skin Barrier. Journal of Investigative Dermatology 131, 1272-1279.

[77] Kirkham PA, Caramori G, Casolari P, et al. Oxidative stress-induced antibodies to carbonyl-modified protein correlate with severity of chronic obstructive pulmonary disease. Am J Respir Crit Care Med. 2011;184:796-802.

[78] Koehler C, Ginzkey C, Friehs G, et al. Ex vivo toxicity of nitrogen dioxide in human nasal epithelium at the WHO defined 1-h limit value. Toxicol Lett. 2011;207:89-95.

[79] Kopp MV, Ulmer C, Ihorst G, et al. Upper airway inflammation in children exposed to ambient ozone and potential signs of adaptation. Eur Respir J. 1999;14:854-61.

[80] Korzun W, Hall J, Sauer R. The effect of ozone on common environmental fungi. Clin Lab Sci. Spring. 2008;21:107-11.

[81] Kosisky SE, Marks MS, Yacovone MA, et al. Determination of ranges for reporting pollen aeroallergen levels in the Washington, DC, metropolitan area. Ann Allergy Asthma Immunol. 2011;107:244-50.

[82] Krishna MT, Springall D, Meng QH et al. Effects of ozone on epithelium and sensory nerves in the bronchial mucosa of healthly humans. Am.J.Respir.Crit.Care Med.1997;156:943-950.

[83] Kuriakose JS, Miller RL. Environmental epigenetics and allergic diseases: recent advances. Clinical & Experimental Allergy. 2010;40:1602–1610.

[84] Kurup VP, Seymour BW, Choi H, Coffman RL. Particulate Aspergillus fumigatus antigens elicit a TH2 response in BALB/c mice. J Allergy Clin Immunol. 1994; 93:1013–20.

[85] Lai ZW, Hundeiker C, Gleichmann E, et al. Cytokine gene expresion during ontogeny in murine thymus on activation of the aryl hydrocarbon receptor by 2,3,7,8-tetrachlorodibenzo-p-dioxin. Mol Pharmacol. 1997;52:30-37.

[86] Leshchuk SI, Popkova SM, Budnikova ZI, et. al. Enteric microbiocenosis in the population of an industrial city. Gig Sanit. 2011; 2:31-5.

[87] Li H, Takizawa A, Azuma et al. Disruption of Nrf2 enhances susceptibility to airway inflammatory responses induced by low-dose diesel exhaust particles in mice, Clin. Immunol. 2008; 128: 366–373.

[88] Li N, Sioutas C, Cho A, et al. Particle air pollutants, oxidative stress and mitochondria damage. Environmental Health Perspectives 2003;111:455-460.

[89] Li N, Sioutas C, Cho A, Schmitz D, et al. Ultrafine particulate pollutants induce oxidative stress and mitochondrial damage. Environmental Health Perspectives. 2003;111:455–460.

[90] Li R, Ning Z, Majumdar R, et al. Ultrafine particles from diesel vehicle emissions at different driving cycles induce differential vascular pro-inflammatory responses: Implication of chemical components and NF-κB signaling. Part Fibre Toxicol. 2010; 7: 6:1-12.

[91] Li Wei Q, Wen Y Xu, Zhu Y Deng, et al. Genome-scale analysis and comparison of gene expression profiles in developing and germinated pollen in Oryza sativa. BMC Genomics. 2010; 11:338.

[92] Li X, Matilainen O, Jin C, et al. Specific SKN-1/Nrf stress responses to perturbations in translation elongation and proteasome activity. PLoS Genet. 2011;7:e1002119.

[93] Liu J, Ballaney M, Al-alem U, et al. Combined inhaled diesel exhaust particles and allergen exposure alter methylation of T helper genes and IgE production in vivo. Toxicol Sci. 2008;102:76-81.

[94] Ma JYC, Ma JKH. The dual effect of the particulate and organic components of diesel exhaust particles on the alteration of pulmonary immune/inflamatory responses and metabolic enzymes.

[95] Majkowska–Wojciechowska B, Pełka J, Balwierz Z, et al. Pollen counts and pevallence of pollen sensitization in children living in rural and urban areas. Allergy Clin Immunol. 2005. Supp.No1:398.

[96] Majkowska–Wojciechowska B, Pełka J, Korzon L, et al. Prevalence of allergy, patterns of allergic sensitization and allergy risk factors in rural and urban children. Allergy 2007;62:1044-1050.

[97] Marklund A, Andersson B, Haglund P. Organophosphorus flame retardants and plasticizers in air from various indoor environments. J Environ Monit. 2005;7:814–819.

[98] Martinez RF. Gene-environment interaction in asthma: with apologies to William of Ockham. Prov Am Thora Soc. 2007:4:26-31.

[99] Martínez-Paz P, Morales M, Martínez-Guitarte JL et al. Characterization of a cytochrome P450 gene (CYP4G) and modulation under different exposures to xenobiotics (tributyltin, nonylphenol, bisphenol A) in Chironomus riparius aquatic larvae. Comp Biochem Physiol C Toxicol Pharmacol. 2011;12. PMID:22019333.

[100] Matthias-Maser, Jaenicke R. The size distribution of primary biological aerosol particles with radii >0.2 μm in an urban/rural influenced region. Atmospheric Research.1995;39:279–286.

[101] Mattila P, Joenväärä S, Renkonen J, et al. Allergy as an epithelial barrier disease. Clinical and Translational Allergy 2011;1:5:1-8.

[102] Mc Lean WH. The allergy gene: how a mutation in a skin protein revealed a link between eczema and asthma. F1000 Med Rep. 2011;14;doi:10.3410/M3-2:1-6.

[103] Mc Creanor J, Cullinan P, Nieuwenhuijsen MJ et al. Respiratory Effects of Exposure to Diesel Traffic in Persons with Asthma. N Engl J Med 2007; 357:2348-58.

[104] Michaeloudes C, Chang PJ, Petrou M, et al. TGF-{beta} and Nrf2 Regulate Antioxidant Responses in Airway Smooth Muscle Cells: Role in Asthma. Am J Respir Crit Care Med. 2011 PMID:21799075.

[105] Miranda ML, Edwards SE, Keating MH, et al. Making the environmental justice grade: the relative burden of air pollution exposure in the United States. Int J Environ Res Public Health. 2011;8:1755-71.

[106] Nemery B, Hoet PH Nemmar A. The Meuse Valley Fog of 1930: an air pollution disaster. Lancet 2001;357:704-8.

[107] Nemmar A, Al-Salam S, Zia S, et al. Diesel exhaust particles in the lung aggravate experimental acute renal failure. Toxicological Sciences. 2009;113:267–277.

[108] Nethery E, Wheeler AJ, Fisher M, et al. Urinary polycyclic aromatic hydrocarbons as a biomarker of exposure to PAHs in air: A pilot study among pregnant woman. J Expo Sci Epidemiol. 2011;14. doi:10.1038/jes.2011.32.

[109] Peden D, Reed CE. Environmental and occupational allergies. J Allergy Clin Immunol. 2010;125:S1: 50-60.

[110] Perera FP, Hemminki K, Grzybowska E, et al. Molecular and genetic damage in humans from environmental pollution in Poland. Nature. 1992;360:256-8.

[111] Peters A, Breitner S, Cyrys J, et al. The influence of improved air quality on mortality risks in Erfurt, Germany. Res Rep Health Eff Inst. 2009; 137:5-77;79-90.

[112] Porębska G, Sadowski M. Contemporary problems of deserts and desertification. Ochrona Środowiska i Zasobów Naturalnych 2007;30.

[113] Rangasamy T, Williams MA, Bauer S, et al. Nuclear erythroid 2 p45-related factor 2 inhibits the maturation of murine dendritic cells by ragweed extract. Am J Respir Cell Mol Biol. 2010;43:276-85.

[114] Renzetti G, Silvestre G, D'Amario C, et al. Less air pollution leeds to rapid reduction airways inflamatory and improved airways function in asthmatic children. Pediatrics 2009; 123:1051-1058.

[115] Rezanejad F. The effect of air pollution on microsporogenesis pollen development and souble pollen proteins in Spartium Junceum L. (Fabiaceae). Turk J Bot. 2007;31:183-91.

[116] Ring J, Eberlin-Koenig B, Bherendt H. Environmental pollution and allergy. Ann Allergy Asthma Immunol. 2001;87:2-6.

[117] Rivera A, Ro G, Van Epps HL et al. Innate immune activation and CD4+ T cell priming during respiratory fungal infection. Immunity 2006; 25:665–75.

[118] Robinson CL, Baumann LM, Romero K, et al. Effect of urbanisation on asthma, allergy and airways inflammation in a developing country setting. Thorax. 2011;thx. 158956, Published Online.

[119] Rückerl R, Greven S, Ljungman P, et al. Air pollution and inflammation (interleukin-6, C-reactive protein, fibrinogen) in myocardial infarction survivors. Environ Health Perspect. 2007;115:1072-80.

[120] Samet JM; Janes H, et al. Fine Particulate Matter and Mortality: A Comparison of the Six Cities and American Cancer Society Cohorts With a Medicare Cohort. Epidemiology. 2008;19:209-216.

[121] Sang N, Yun Y, Yao GY, et al. SO2-induced neurotoxicity is mediated by cyclooxygenases-2-derived prostaglandin E2 and its downstream signaling pathway in rat hippocampal neurons. Toxicol Sci. doi: 10.1093/toxsci/kfr224.

[122] Schelegle ES, Miller LA, Gershwin LJ, et al. Repeated episodes of ozone inhalation amplifies the effects of alergen sensitization and ihalation on Airways immune and structural development In Rhesus monkeys. Toxicology and Applied Pharmacology. 2003;191:74-85.

[123] Singer BD, Ziska LH, Frenz DA, et al. Increasing Amb a 1 content in common ragweed (Ambrosia artemisiifolia) pollen as a function of rising atmospheric CO2 concentration. Functional Plant Biology. 2005;32:667–670.

[124] Somssich IE, Schmeizer E, Bollman J, et al. Rapid activation by fungal elictor of genes encoding „pathogenesis-related" proteins In cultured parsley cells. Proc Nati Acad Sci USA. 1986;83:2427-2430.

[125] Stapleton HM, Klosterhaus S, Keller A, et al. Identification of flame retardants in polyurethane foam collected from baby products. Environ Sci Technol. 2011;15:5323-31.

[126] Stępkowski TM, Kruszewski MK. Molecular cross-talk between the NRF2/KEAP1 signaling pathway, autophagy, and apoptosis. Free Radic Biol Med. 2011 1;50:1186-95.

[127] Suarez-Cervera M, Castells T, Vega-Maray A, et al. Effects of air pollution on cup a 3 allergen in Cupressus arizonica pollen grains. Ann Allergy Asthma Immunol. 2008;101:57-66.

[128] Sulentic CE, Kaminski NE. The long winding road toward understanding the molecular mechanisms for B-cell suppression by 2,3,7,8-tetrachlorodibenzo-p-dioxin. Toxicol Sci. 2011;120 Suppl 1:S171-91.

[129] Supanc V, Biloglav Z, Kes VB et al. Role of cell adhesion molecules in acute ischemic storke. Ann Saudi Med. 2011;31:365-70.

[130] Sussan TE, Rangasamy T, Blake DJ, et al. Targeting Nrf2 with the triterpenoid CDDO-imidazolide attenuates cigarette smoke-induced emphysema and cardiac dysfunction in mice. Proc Natl Acad Sci 2009;106:250–255.

[131] Szabo SJ, Sullivan BM, Stemmann C, et al. Distinct effects of T-bet in TH1 lineage commitment and IFN-gamma production in CD4 and CD8 T cells. Science. 2002; 295:338-42.

[132] Takenaka H, Zhang K, Diaz-Sanches D, et al. Enhanced human IgE production results from exposure to the aromatic hydrocarbons from diesel exhaust: direct effects on B-cell IgE production, J Allergy Clin Immunol. 1995:95:103-115.

[133] Takeshita A, Igarashi-Migitaka J, Nishiyama K, et al. Acetyl tributyl citrate, the most widely used phthalate substitute plasticizer, indu ces cytochrome P450 3A through steroid and xenobiotic receptor. Toxicol Sci. 2011;123:460-70.

[134] Tarkowski M, Kur B, Nocun M, et al. Perinatal exposure of mice to TCDD decreases allergic sensitisation through inhibition of IL-4 production rather than T regulatory cell-mediated suppresion. International Journal of Occupational Medicine and Environmental Health. 2010; 23:75-84.

[135] Tsien A, Fleming J, Saxon A. Combined diesel exhaust particulate and ragweed allergen challenge markedly enhances human in vivo nasal ragweed-specific IgE and skews cytokine production to a t helper cell 2-type pattern. J Immunol. 1997;158:2406–2413.

[136] Uysal N, Schapira RM. Effects of ozone on lung function and lung diseases. Curr Opin Pulm Med. 2003;9:144-150.

[137] Van Loon LC, Pierpoint WS, Boller Th et al. Recomendations for naming plant pathogenesis-related proteins. Plant Mol Biol. Report 1994;12:245-246.

[138] Van Loon LC, Van Strien EA, The familie of pathogenesis-related proteins, their activities, and comparitive analysis of PR-1 type proteins. Physiol Mol Plant Pathol. 1999;55:85-97.

[139] Volling K, Thywissen A, Brakhage AA, et al. Cell Microbiol. Phagocytosis of melanized Aspergillus conidia by macrophages exerts cytoprotective effects by sustained PI3K/Akt signalling. 2011;13:1130-48.

[140] Wang J, Liu X Li T, et al. Increased hepatic Igf2 gene expression involves C/EBPβ in TCDD-induced teratogenesis in rats. Reprod Toxicol. 2011;32:313-21.

[141] Whangchai K, Saengnil K, Uthaibutra J. Effect of ozone in combination with some organic acids on the control of postharvest decay and pericarp browning of longan fruit. Crop Protection. 2006;25:821-825.

[142] Wichmann G, Frank U, Herbarth O, et al. Toxicology 2009; 257:127-136.

[143] Wierda D, Irons RD, Greenlee WF, Immunotoxicity In C57B1/6 mice expose to benzene and Arclor 1254. Toxicol. Appl. Pharmacol.1981, 60,410-417.

[144] Williams MA, Rangasamy T, Bauer S, et al. Disruption of the Transcription Factor Nrf2 Promotes Pro-Oxidative Dendritic Cells That Stimulate Th2-Like Immunoresponsiveness upon Activation by Ambient Particulate Matter . The Journal of Immunology, 2008, 181, 4545-4559.

[145] Wold LE, Simkhovich BZ, Kleinman MT, Nordlie MA, Dow JS, Sioutas C, Kloner RA. In vivo and in vitro models to test the hypothesis of particle-induced effects on cardiac function and arrhythmias. Cardiovasc Toxicol. 2006;6:69–78.

[146] Wu K, Xu X, Liu J, et al. In utero exposure to polychlorinated biphenyls and reduced neonatal physiological development from Guiyu, China. Ecotoxicol Environ Saf. 2011. 2011;74:2141-7.

[147] Wudarczyk A. Toksyczny wpływ tlenków azotu na organizm człowieka. 2001; publ. online.

[148] Yanagisawa R, Takano H, Inoue K et al. Components of diesel Exhaust particles differentially affect Th1/Th2 response In a Marine model of allergic inflamation. Clin Exp Allergy 2006;36:386-395.

[149] Yin XJ, Ma JY, Antonini JM et al. Roles of reactive oxygen specjes and heme oxygenase-1 in modulation of alveolar macrophage-mediated pulmonary immune responses to listeria monocytogenes by diesel Exhaust particles. Toxicological Sciences 2004; 82:143-153.

[150] Zhou L, Littman DR. Transcriptional regulatory networks in Th17 cell differentiation. Curr Opin Immunol. 2009; 21: 146–152.

[151] Zhu H, Han J, Xiao JQ, et al. Uptake, translocation and acumulation of manufactured iron oxide nanoparticles by pumpkin plants. Journal of Environmental Monitoring. 2008;10:713-717.

[152] Ziska L, Knowlton K, Rogers C, et all. Recent warming by latitude associated with increased length of ragweed pollen season in central North America. Proc Natl Acad Sci U S A. 2011;108: 4248–4251.

10

Design Efficiency of ESP

Maria Jędrusik and Arkadiusz Świerczok
Wroclaw University of Technology
Poland

1. Introduction

Electrostatic precipitator (ESP) is a highly efficient device for cleaning exhaust gases from industrial processes. Fig. 1 shows a photograph of electrostatic precipitator (ESP) for cleaning flue gases from a power boiler, and Fig. 2 shows the inner part of the ESP.

The basic components of ESP are: a chamber comprising discharge and collection electrodes, and high voltage (HV) supply unit - shown in Fig. 3. The ESP operation is based on the utilization of the influence of electric field on charged dust particles flowing between the electrodes. The Discharge electrodes (DE) are connected to a direct current (DC) HV power supply of negative polarity and the collecting electrodes to the positive pole of the supply, which is additionally grounded. The raw exhaust gas is subjected to electron and ion currents which charge the dust particles negatively, and cause their movement towards the collecting electrode. The precipitation of dust particles from a gas stream as well as its collection occurs mainly because of the electrophoresis forces.

Fig. 1. Electrostatic precipitator in Thermal-Power Station.

Fig. 2. Inner part of the ESP.

Dust particles collected on the CE surfaces partly give up their charge and the dust layer is kept on the CE electrodes by means of mechanical and electrical forces. Afterwards the collected dust layer is knocked down mostly by mechanical rapping systems.

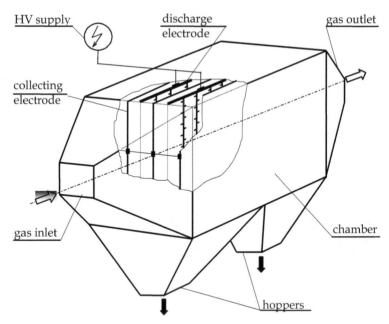

Fig. 3. Typical arrangement of wire-and-plate precipitator with horizontal gas flow.

2. Kinetics of dust particle charging

The dust particles in ESP are charged as a result of taking over the electric charge from gaseous ions. The source of gas ionization is a negative corona discharge originating at the DEs.

The discharge takes place due to strongly inhomogeneous electric field in the vicinity of appropriately formed DE surface, for example, in the form of thin round wire or a similar element with mounted spikes. The empirical equation of the corona-onset electric field strength on DE spikes has been given by Peek (Peek, 1929).

$$E_o = A\delta + B\sqrt{\frac{\delta}{r_o}} \tag{1}$$

where:

- E_o – initial electric field strength, V/m
- A, B – experimental coefficients characterizing gas type and discharge polarity. For an ESP with negative discharge polarity Robinson (Parker, 1997) advices to use empirical values: $A=3.2 \cdot 10^6$ V/m, $B=9 \cdot 10^4$ V/m$^{1/2}$,
- r_o – DE curvature radius, m
- δ – relative density of gas, -

The magnitude of supply voltage at which the corona discharge begins on the DE surface is called the corona onset voltage. Above this level, develop the electron avalanches from the discharge electrode towards the plate. The electrons emitted from the spikes are accelerated in the strong electric field and gain energy necessary for avalanche ionization of atoms and gaseous molecules. Additional source of electrons in the discharge is also the so called secondary emission due to positive ions impacting the DE. The avalanches originating from DE develop in the direction of CE. Electrons from the avalanche head are quickly attach to neutral gas molecules , which-become negative gas ions. Dust particles get electric charge due to non-elastic collisions with negative as well as positive gas ions. In the charging process of dust particles, two distinguish basic mechanisms are considered (White, 1990):

- field charging,
- diffusion charging.

An equation describing the charge on a dielectric spherical particle for the field charging mechanism has been given by Pauthenier and Moreau-Hanot (Pauthenier & Moreau-Hanot, 1932) in the following form:

$$q_f = q_s \cdot \frac{t}{t+\tau} \tag{2}$$

where: q_f – particle charge obtained from field charging,

$$q_s = \pi\varepsilon_o \frac{3\varepsilon_w}{\varepsilon_w + 2} Ed^2 \quad - \quad \text{particle saturation charge,}$$

$$\tau_f = \frac{4\varepsilon_o E}{j} \quad - \quad \text{field charging constant,}$$

- ε_w – dielectric constant of particle material,
- ε_o – dielectric constant of free space $\left(8.85\cdot 10^{-12}\ \dfrac{C^2}{Nm^2}\right)$
- E – electric field strength,
- t – charging time,
- d – particle diameter.

White in 1963 (White, 1990) has given the equation of particle charging for diffusion mechanism in the form of:

$$q_d = \frac{2\pi\,\varepsilon_o kTd}{e}\ln\left(1+\frac{t}{\tau_d}\right) \tag{3}$$

where:

- q_d – particle charge obtained from diffusion charging
- k –Boltzmann constant ($1.38\cdot 10^{-23}$ J/K)
- T – temperature
- e – electron charge, ($e = 1.67\cdot10^{-19}$ C)
- τ_d – diffusion charging constant

$$\tau_d = \frac{\varepsilon_o\sqrt{8m_j k\pi T}}{e^2 Nd} \tag{4}$$

- N – number of unipolar ions in the unit volume (ion density)
- m –mass of an ion

It was experimentally demonstrated that a total charge of dust particle can calculated with practically sufficient accuracy as a sum of field charge (2) and diffusion charge (3).

$$q_p(t) = q_f(t) + q_d(t) \tag{5}$$

It should be noted that all of the above mentioned charging theories apply only to spherical particles. When taking into account industrial dust particles it is necessary to use their equivalent dimensions (diameter).

In typical industrial ESP, the dust size distribution at the precipitator inlet does not comprise fine particles –below 4 µm, and the electric field is usually over 1 kV/m. Therefore, based on the research results presented in the literature, it is generally accepted that the mechanism of diffusion charging may be ignored. This also proves that the Pauthenier's & Moreau-Hanot equation describes the kinetics of dust particle charging with sufficient accuracy.

3. Dust particle motion in ESP

In order to characterize the movement of charged particle in an ESP it is necessary to assume the equilibrium of forces acting on the particle. After some simplifications it can be said that the following forces are acting on a dust particle in an ESP: the inertia force, electric force and drag force of the medium, where:

$$\vec{F}_i = -m \cdot \frac{d\vec{u}}{dt} \tag{6}$$

$$\vec{F}_e = q_s \cdot \vec{E} \tag{7}$$

$$\vec{F}_d = c_d(\mathrm{Re}) \cdot \rho_g \cdot \frac{\pi d}{8}(\vec{v} - \vec{u})|\vec{v} - \vec{u}| \tag{8}$$

The motion of any dust particle may be described as by the Newton second law:

$$\vec{F}_i + \vec{F}_e + \vec{F}_d = 0 \tag{9}$$

The scheme of particle motion in electric field illustrates Fig. 4.

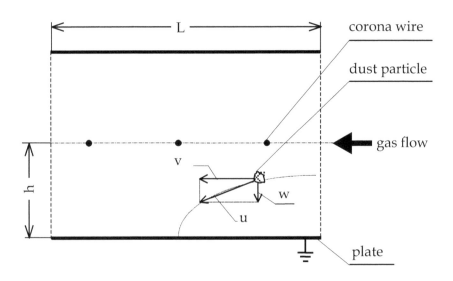

Fig. 4. Schematic diagram of particle motion in electric field (plate to plate configuration).

In the vector form, the equation of motion becomes:

$$m\frac{d\vec{u}}{dt} = -c_d(\mathrm{Re})\,\rho_g\,\frac{\pi d}{8}(\vec{v} - \vec{u})|\vec{v} - \vec{u}| + q_s\vec{E} \tag{10}$$

where:

- $\vec{v}$ - vector of gas velocity
- $\vec{u}$ - vector of particle velocity
- ρ_g - gas density
- $\vec{v} - \vec{u}$ - particle velocity in relation to gas velocity
- $c_d(Re)$ – dynamic drag coefficient
- m – particle mass

It should be emphasized that the dominant role in this equation play the electric force and drag force of the gas medium. In the steady state motion the inertia force can be omitted because of its low value comparing to the electric force (Parker, 1997).

The equation (10) finally gets the form:

$$-c_d(\text{Re}) \cdot \rho_g \frac{\pi d}{8}(\vec{v} - \vec{u})|\vec{v} - \vec{u}| + q_s \vec{E} = 0 \tag{11}$$

3.1 Theoretical migration velocity

Accepting for further consideration the simplest case of spherical particle steady motion in electric field –in the range of Stoke's law (Re ≤ 0.1), equation (11) can be transformed to the form:

$$q_s E - 3\pi\mu d \cdot w = 0 \tag{12}$$

where: w – particle relative velocity normal to CE surface; so called migration velocity $w = (\vec{v} - \vec{u})$

Theoretical value of the migration velocity, calculated from equation (12) equals to:

$$w = \frac{q_s \cdot E}{3\pi\mu d} \tag{13}$$

The minimum range of the size of particles to which the Stoke's equation can be applied is the case when the particle diameter is of the order of magnitude of mean free path of gas molecules $\bar{\lambda}$. For particles smaller than 1 μm it is necessary to take into account the Cunningham slip correction factor:

$$w = \frac{q_s E C_u}{3\pi\mu d} \tag{14}$$

where:

C_u - Cunningham slip correction factor (White, 1990); $C_u = 1 + 0.86 \frac{2\bar{\lambda}}{d}$

The formulas used for the calculation of theoretical migration velocity do not take into account many factors affecting the movement of dust particle in electric field such as:

inertia, inhomogeneity of electric field strength distribution, gas velocity, and electric wind velocity.

In the electrostatic precipitation process with a spike and plate electrodes arrangement there exists an electro-hydro-dynamic (EHD) flow, which is an effect of mutual interaction of electrically neutral main gas stream and gas ions movement under the influence of electric field. To describe such flow field it is necessary to use dimensionless parameters determined by IEEE-DEIS-EHD Technical Commitee (IEEE-DEIS-EHD TC, 2003).

$$\text{Re} = \frac{L \cdot \bar{v}}{v} \tag{15}$$

$$Ehd = \frac{L^3 \cdot j_0}{v \cdot \rho_g \cdot b \cdot A} \tag{16}$$

$$Md = \frac{\varepsilon_0 \cdot E_0^2 \cdot L^2}{\rho_g \cdot v} \tag{17}$$

where:

- L – characteristic length, i.e. distance between the electrodes, m
- j_0 –total discharge current, A
- $\bar{v}$ - average gas velocity, m/s
- v – kinematic viscosity coefficient, m²/s
- Md – the Masuda number,
- Re – the Reynolds number,
- b – ions mobility: $1.8 \cdot 10^{-4}$, m²/Vs
- Ehd – electro-hydro-dynamic (EHD) number,
- E_0 – field strength at corona onset, V/m
- ρ_g – gas viscosity, kg/m³
- A – CE surface for discharge current calculations, m²

For specified ESP arrangement, the Reynolds number depends on gas flow velocity, and the EHD and Masuda numbers are the functions of discharge voltage, field geometry, and ionization parameters of the gas.

3.2 ESP precipitation efficiency

The basic equation describing precipitation efficiency from the probability theory has been given in 50-ties by White (White, 1990) and latter modified by Matts & Oehnfeld to the following form:

$$\eta(d) = 1 - \exp\left\{-w_t(d) \cdot \frac{L}{h \cdot v}\right\} \tag{18}$$

where:

- $\eta(d)$ – precipitation efficiency for a particle with diameter d,

-   $w_t(d)$ –theoretical migration velocity, m/s
-   L - length of electric field, m
-   h - wire-plate distance, m

The total precipitation efficiency $\eta_C(d)$ can be calculated from the formula:

$$\eta_C(d) = \sum_{d_{min}}^{d_{max}} q_3(d)\eta(d) \tag{19}$$

Often an alternative way to determine the total efficiency of precipitation is calculate it by measuring the dust concentration before and after ESP.

4. The influence of combustion process and fired coal parameters on physical & chemical properties of generated fly ash

4.1 Chemical composition of fly ash

The fly ash collected in an ESP is a mixture of different compounds, mainly of silicon and aluminum oxides with average substitute diameter of about 15 µm and submicron particles with diameter below 1 µm (ca. 2wt.%). Characteristic properties of fly ash having the greatest influence on ESP operation are (Parker, 1997): diameter, form and structure of particles, their propensity for agglomeration and cohesion, electrical resistivity, chemical composition and reactivity. The chemical composition of fly ash allows to estimate its predictable electrical resistivity value and, by this way, the required size of the ESP (Chambers et al., 2001). Actually, it often becomes necessary to adapt an existing ESP to new (changed) operational conditions, for example, after installing flue gas desulfurization equipment (Parker, 1997). Also the installation of low-emission burners in boiler results in increasing amount of combustible elements in fly ash (LOI). In that case takes place, changes of the gas-dust medium parameters as well as its electric resistivity can be expected. Former experiences with the electrostatic gas cleaning process led to the conclusion that the dust electrical resistivity is an important parameter influencing the operational efficiency of ESPs. If the dust electric resistivity exceeds 10^{11}–10^{13} $\Omega\cdot$cm it is the so called high resistivity dust, which is difficult to collect. If the resistivity lies between 10^{10}–10^{11} $\Omega\cdot$cm, it is in the optimal range for the collection. The chemical composition of fly ash is closely related to the coal quality. An increase of silicon and alumina compounds in the fly ash (SiO_2, Al_2O_3) may lead to the increase of fly ash electric resistivity and by that to decrease of the ESP collection efficiency. It has been observed that for brown coal fly ash, the electric resistivity increases as the percentage of alkali compounds (CaO + MgO) exceeds 3-6 times that of iron trioxide (Fe_2O_3). However, significant amount of sodium and potassium compounds in fly ash cause a decrease in its resistivity that is particularly noticeable by high content of (SiO_2 + Al_2O_3) (Bibbo, 1994; Bickelhaupt, 1985; Parker, 1997).

A substantial influence on the fly ash electric resistivity has the content of sulphur trioxide in the flue gas entering into ESP, as shown on Fig 5. When firing coal in a combustion chamber, the sulphur contained in the coal is oxidized to SO_2. Depending on the combustion conditions, 0.5-3% of that sulphur dioxide is further oxidized to SO_3. At the temperature of sulphur acid dew point, the condensation of SO_3 on particle surfaces takes place – or more precisely- H_2SO_4 is formed on it in the form of very thin film.

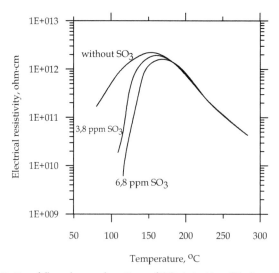

Fig. 5. Electric resistivity of fly ash as a function of SO_3 injection (Parker, 1997).

In Table 1 are presented selected characteristic parameters of fly ash resulting from combustion of hard coal and brown coal in different power boilers.

Chemical composition, %	Fly ash type							
	E	D	B	C	W	A	JG	G
SiO_2	54.0	41.00	41.60	37.60	45.67	54.20	47.44	28.99
Fe_2O_3	4.21	10.30	6.50	7.01	8.94	5.30	6.91	3.67
Al_2O_3	4.42	30.60	21.90	21.60	21.65	32.10	19.65	17.14
TiO_2	1.03	2.08	0.85	0.81	1.09	1.40	0.99	0.86
CaO	25.90	3.03	11.90	14.30	8.23	0.81	3.98	2.82
MgO	4.43	1.97	2.29	2.47	2.60	1.09	1.41	1.01
SO_3	4.72	2.80	6.27	6.58	1.57	0.27	0.73	2.26
K_2O	0.24	1.28	2.24	1.87	4.83	2.65	3.03	2.68
P_2O_5	0.26	0.22	0.15	0.16	-	0.55	0.01	0.01
Na_2O	0.09	3.61	1.22	1.48	1.32	0.48	1.33	1.14
Un-burned coal	0.63	0.14	2.38	2.68	3.50	0.61	13.77	28.60
Density, kg/m^3	2500	1954	2580	2690	2210	2031	1550	1580
Resistivity, $\Omega \cdot cm$	4.4×10^8	3.2×10^7	2.0×10^8	1.8×10^8	3.2×10^7	1.8×10^8	5.1×10^7	5.0×10^7

Table 1. Properties of fly ashes.

The influence on chemical composition of fly ash have the quality of fired coal and the combustion parameters. Because both of the mentioned parameters vary with time, the chemical composition of fly ash is also changed with time.

4.2 Dust particle size distribution

Knowledge of the particle size (granulation) distribution is essential to estimate an ESP collection efficiency. The fly ashes coming after combustion of solid fuels are polydisperse and diameter of the particles ranges from fractions of micrometer up to several millimeters. Determination of particles size is a difficult task because of various shapes of the particles, from spherical forms -created as an effect of sublimation and condensation, spatially expanded, inside-empty structures of un-burned coal, snow-flake like flat particles, to fibrous particles. In order to compare the dust size distributions, a equivalent particle diameter has been introduced. It depends on the method of size analysis: the projected diameter (determined by the analysis under projecting microscope or by sieve analysis) or dynamic diameter (obtained using the blow away method in counter-flow, or sedimentation).

The fly ash size distribution is most often presented as fraction of particles $q_r(d_i)$ in a range from d_i to $d_i + dd$, or the total number of particles $Q_r(d_i)$ smaller than d_i (cumulative size distribution). The particle distribution in a certain size range may be represented by its mass, volume or number ratios. These ratios are called the mass, volumetric and number fractions with the index r equal to 3 (mass and vol.) or 0 (number), respectively (Masuda et al., 2006).

Examples of fly ash size distribution coming from different boilers fired with hard or brown coal are presented in Figs. no 6, 7 & 8. The analysis has been done with an automatic particle size analyzer *Mastersizer S* made by *Malvern Instruments Ltd*. Results of the presented analyses show that the combustor type (boiler type) is a crucial element in forming the fly ash size distribution character.

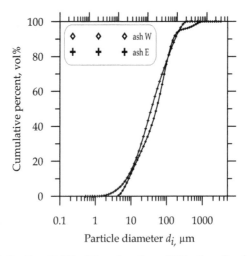

Fig. 6. Particle size distribution $Q_3(d_i)$ of fly ashes from PC boilers fired with hard coal (fly ash W) and brown coal (fly ash E).

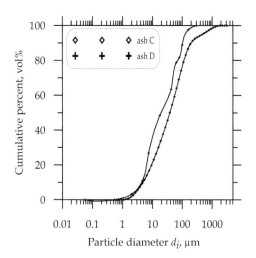

Fig. 7. Particle size distribution $Q_3(d_i)$ of fly ashes from PFB boilers fired with hard coal (fly ash C) and brown coal (fly ash D).

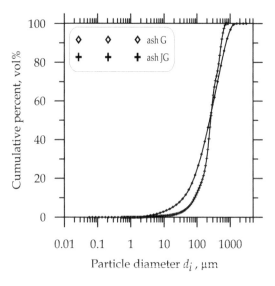

Fig. 8. Particle size distribution $Q_3(d_i)$ of fly ashes from grate stoker boilers fired with hard coal.

The influence of ESP device on fly ash size distribution is presented in Fig. 9. At the outlet of a high efficiency ESP ($\eta_C > 99.9\ \%$), the fly ash comprises mainly of fine particles having diameter below 20 μm.

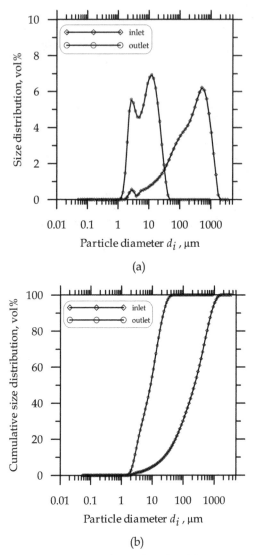

Fig. 9. Particle size distribution at an ESP inlet and outlet: (a) particles fraction $q_r(d_i)$, (b) cumulative size distribution $Q_3(d_i)$, fly ash from grate stoker boiler fired with hard coal.

4.3 Particle forms of a fly ash

Different methods of measuring fly ash size distribution utilize the same geometric parameter i.e. particle substitute diameter. But as it was mentioned before, the actual shape (form) of particles are rare spherical that also influences their separation process in an ESP.

In Fig. 10 are shown different particle-shape patterns, which can be found in various fly ashes: spherical forms (spherules) and sharp-edged (Fig. 10a), particles in the form of fibers and particles with a very irregular shapes (Fig. 10b). Moreover there is also visible a significant particle size diversification. Scanning Electron Microscope (SEM) micrographs taken at high magnification show the complexity of the forms of particles, which are often agglomerates of many smaller particles having different diameters.

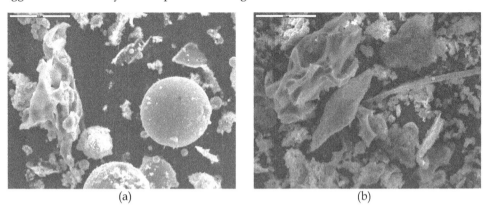

(a)   (b)

Fig. 10. SEM pictures of fly ash particles from hard coal fired boilers: (a) in a grate stoker boiler (fly ash G), and in a PC boiler (fly ash C) (magnification 700x).

On the photo (Fig. 11) are shown characteristic shapes (forms) of fly ash particles coming from brown coal fired boilers with different combustor systems.

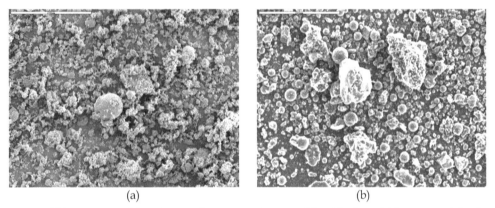

(a)   (b)

Fig. 11. SEM pictures of fly ash particles coming from coal fired boilers: (a) brown coal fired in fluidal bed boiler (fly ash D), (b) brown coal fired in PC boiler (fly ash E) (magn. 230x).

The elemental analysis carried out by Energy Dispersive X-Ray spectroscopy (EDX) method demonstrates that most of the particles are alumina-silica (Al_2O_3-SiO_2) aggregates (Fig. 12) as well as spherical granules of two kinds: built of alumina-silica and spherical forms of iron oxides (Fig 13). In addition to that in the fly ash were found particles with compounds characteristic of carbonates, sulfates and oxides (quartz, feldspar, calcite and gypsum), with considerable addition of titanium, iron, potassium, calcium, plus small content of sulfur and potassium (Grafender, 2010).

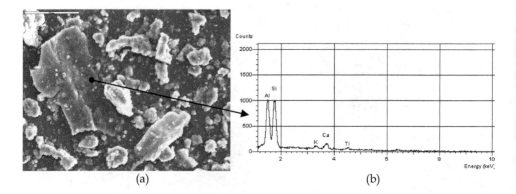

(a)   (b)

Fig. 12. Fly ash particles composed of alumina-silicates (magn. 700x) - (a) and their elemental analysis - (b), fly ash from CFB boiler fired with brown coal (fly ash D).

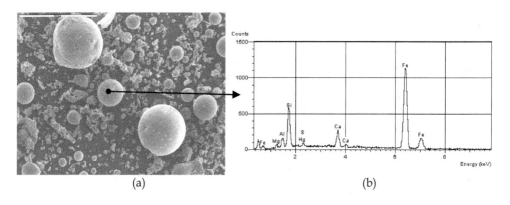

(a)   (b)

Fig. 13. Fly ash particles of spherical form composed of iron-oxides (magn. 100x) - (a) and their elemental analysis - (b), fly ash from grate stoker boiler fired with hard coal (fly ash JG).

5. Discharge Electrode (DE) model investigation

5.1 Testing bench

The model investigations of discharge electrodes (DE) have been carried out in a laboratory arrangement comprised of pilot ESP with horizontal air flow, as shown on Fig. 14. The chamber is made of organic glass (2000 mm long, 400 mm wide and 450 mm high) that enables visual observations as well as photography of the phenomenon occurring in the inter electrode region.. Tests were carried out with air flow at a temperature of 20°C , pressure 1000 hPa and at humidity of 60% (Jędrusik & Świerczok, 2009).

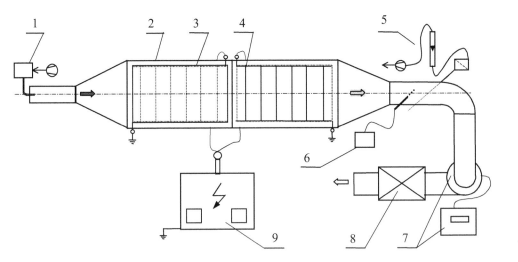

Fig. 14. Laboratory arrangement for DE testing in a pilot ESP : 1 – fly ash feeder, 2 – pilot ESP chamber, 3 – CE (collecting electrodes), 4 – DE (discharge electrodes), 5 – dust meter, 6 - thermo anemometer , 7 – exhaust fan with rotational speed control, 8 – final filter, 9 – HV (high voltage) supply unit.

5.2 V-I (voltage-current) characteristics

In Fig. 15 are shown various constructions of tested rigid discharge electrodes (RDE). In electrodes of this type, both functions of the construction: mechanical supporting and electric-discharge generation have been separated via mounting the active spikes as replaceable elements that allows replacement of the emission points without changing the supporting part. The V-I characteristics shown in Fig. 16 allows to divide the considered RDE constructions into two groups:

1. 'aggressive' (with steep V-I curve) – the so called 'RDE-3', having discharge onset at a level of U_0=10 kV and the 'barbed type' with higher onset voltage of about U_0=22 kV; and
2. 'smooth': RDE-1 with discharge onset level of U_0=16 kV and RDE-2 with U_0=14 kV (Jędrusik & Świerczok, 2011).

The tests have shown that modification of spikes orientation and spacing influences the V-I curvature, what can be seen in Fig. 17. That gives the possibility to select and optimize DE electrodes according to required precipitation efficiency and the expected shape of its V-I characteristic. This becomes important when fly ash parameters are changed (mainly its resistivity), for example, as a result of changing the kind fired fuel. Hitherto existing experience shows that for efficient precipitation of high-resistivity fly ash the DE construction should allow a high discharge voltage and uniform discharge current distribution. Such electrode is called " high voltage & moderate discharge current electrode".

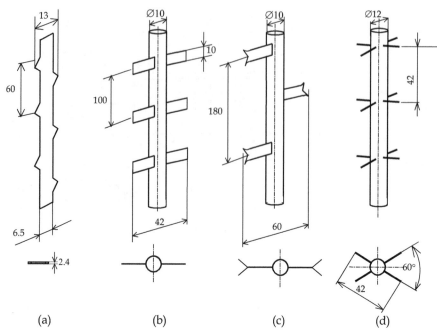

Fig. 15. Forms of discharge electrodes (DE): (a) 'barbed tape', (b) RDE–1, (c) RDE–2, (d) RDE–3.

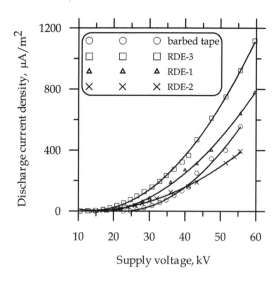

Fig. 16. V-I characteristics of DE electrodes shown on Fig. 15.

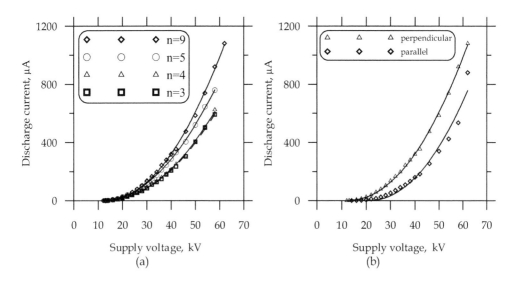

Fig. 17. V-I characteristics of RDE-2 electrode: (a) effect of discharge spikes number, (b) normal and parallel orientation of the spikes to the collection electrode CE.

At the end of 90s (of the 20th century) there were carried out many laboratory tests with various constructions of DE as well as with numerical modeling of phenomenon occurring in an electric discharge field regions for different 'spikes' of the electrodes (Brocilo et al., 2001; Caron & Dascalescu, 2004; Chung-Liang & Hsunling, 1999; Hsunling et al., 1994; McCain, 2001).

Regardless of those investigations, there still lack unambiguous criteria for the selection from various available constructions of DE. Very often ESPs are equipped with similar type of DEs irrespective on the gas-dust characteristic parameters or inter electrode spacing.

5.3 The influence of selected fly ash parameters on precipitation efficiency

In order to show the influence of fly ash chemical composition on precipitation efficiency a several measurements were done on a pilot ESP with selected fly ashes (parameters presented in Table 1) and selected DE constructions.

To illustrate the results, in Fig. 18 are shown characteristics of precipitation efficiency for three different fly ashes. The curves demonstrate that high content of compounds like Al_2O_3 (32.1%), SiO_2 (54.2%) with traces of SO_3 Na_2O in the fly ash decreases the ESP precipitation and efficiency -at the specific experiment conditions. For example, a 10% decrease of Al_2O_3 in the fly ash and increase of SO_3 up to 6% and Na_2O up to 1% cause an increase in the precipitation efficiency of fly ash that may indicate a favorable influence of sodium content in the fly ash (Jędrusik & Świerczok, 2006; Jędrusik, 2008).

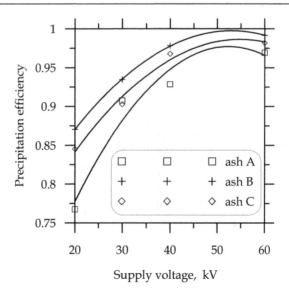

Fig. 18. Precipitation efficiency vs. supply voltage for RDE-2 electrode.

There was also tested the influence of unburned coal (LOI) content in fly ash on the precipitation efficiency, and an example of experimental results are presented in Fig. 19.

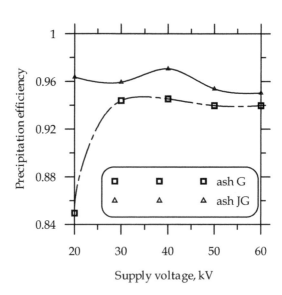

Fig. 19. Precipitation efficiency as a function of supply voltage for RDE-3 electrode and fly ash from hard coal fired grate stoker boiler.

The characteristics shown in Fig. 19 present the influence of unburned coal content in fly ash on the precipitation efficiency that was already observed in research works in 70'th of the 20th century. An increase of unburned coal percentage by over 15% decreased the precipitation efficiency (Hagemman & Ahland, 1973).

There was also tested the influence of biomass (of plant origin) co-firing in power boilers on precipitation process in the ESP, what is shown in Fig. 20.

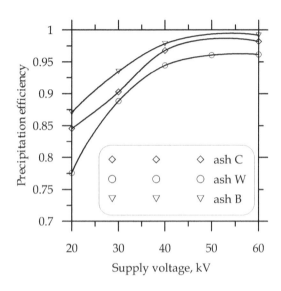

Fig. 20. The relationship between precipitation efficiency and a biomass percentage in the co-firing process (hard coal), RDE-2 electrode.

It is shown that the collection efficiency depends on electrical parameters of the supply voltage and the biomass percentage. The collection efficiency increases with an increase of the supply voltage of the discharge electrode, but it is saturated for a certain voltage magnitude, of about 50 kV, for that specific case. Further increase of the voltage can even cause a slight decrease of collection efficiency. It was also determined, that small addition of biomass (10%) to bituminous coal (ash B) causes an increase of the collection efficiency, whereas for higher content of biomass, 50% (ash W) or larger, the collection efficiency decreases. These preliminary results indicate that further research on the effect of co-fired biomass content on the collection efficiency is required in order to optimize the operational parameters of electrostatic precipitator (Jaworek et al, 2011).

The optimization of DE (corona electrode) design should include not only the parameters of the electric field, but also the physical and chemical properties of the fly ash. In summary, the choice of an appropriate design of the discharge electrode should be based on a thorough examination of the dust particles and flue gas properties.

5.4 Current density distribution and patterns of precipitated dust on (CE) electrodes

The measurements of current density enables better estimation of selected DE constructions especially in connection with local accumulation of fly ash on CE surfaces. The deposition of dust in an ESP creates collection patterns, which shape depends on the electric field forces in the inter-electrode space (Miller et al. 1996a, 1996b).

A measuring arrangement diagram is shown on Fig. 21.

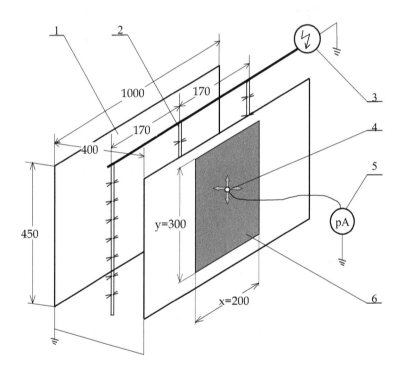

Fig. 21. Measuring arrangement of discharge current distribution on the CE surface:
1 – collecting electrodes, 2 – discharge electrodes, 3 – HV supply unit, 4 –measuring panel,
5 – pico-ammeter, 6 – measuring zone.

In Fig. 22 is presented discharge current distribution for RDE-3 electrode (Fig. 15d) with 'spikes' pointed perpendicularly at the surface of CE. In Fig. 23 is shown pattern of collected fly ash on CE electrodes for this DE construction.

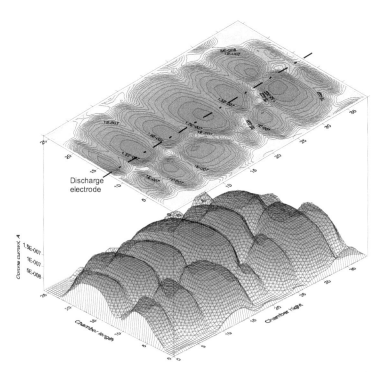

Fig. 22. Discharge current distribution for RDE-3 electrode - supply voltage 50kV.

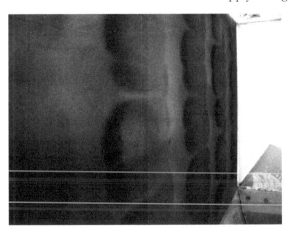

Fig. 23. Pattern of collected fly ash on CE electrodes for RDE-3 electrode.

From the results appears that the highest value of discharge current density is opposite the DE 'spike'. Hence the uniformity of discharge current distribution, which is important for high precipitation efficiency, will depend on the DE 'spikes' number and their configuration (Blanchard et al., 2002, McKinney et al., 1992). For this reason it is crucial to use DE constructions, which limit the number and area of regions with very high or very low current density. Observation of the collection patterns on CE surfaces enables qualitative assessment of the discharge current distribution on the electrode. There is visible a significant correlation between the collected fly ash patterns and the measured distribution of discharge current. From the studies presented in (Miller et al., 1996a) also results that the collected fly ash layer density depends on the collection pattern, in which the highest density of the fly ash layer appears opposite the DE 'spikes', that should be related to the electric field distribution between the electrodes. This phenomenon may also be utilized in designing and selecting DE for collection of submicron particles.

6. Summary

The results presented in this Chapter have shown that different constructions of RDE electrodes in ESPs, their 'spike' number and geometrical configuration have to be used depending on physical and chemical properties of fly ash. Although the model studies have been carried out for only a few types of DE constructions and selected kinds of fly ashes, the experimental results, confirmed by the literature's data, had shown the influence of fly ash chemical composition as well as DE construction on the total collection efficiency of ESP. It was confirmed that some components of fly ash (e.g. Na_2O or Al_2O_3) have different effect on the collection efficiency, depending on DE construction and the type of fired coal (hard or brown coal). The results of measurements obtained for selected DE constructions in a pilot ESP have shown that the construction of DE, i.e., shape of their 'spikes', number of spikes, and their orientation relative to the collection electrode, have an influence on voltage-current characteristics and the corona onset voltage. These results suggest that voltage-current characteristics can be changed to some extent via changing the discharge electrode geometry (shape of spikes) or the modification of electrodes configuration. The possibility of the formation of V-I characteristics adequately to the existing collection conditions, enables more effective exploitation of H.V. supply units, in order to get higher collection efficiency of ESP and increasing energy efficiency of the supply unit. The presented results indicate also on new possibilities of more efficient removal of submicron particles in industrial ESPs.

7. References

Bibbo P.P. (1994). Agential flu gas conditioning for electrostatic precipitator, *Proc. of the American Power Conference*, Illinois Institute of Technology, USA, Vol. 56/V1,1994

Bickelhaupt R.E. (1985). A study to improve a technique for predicting fly ash resistivity with emphasis on the effect of sulfur trioxide, Prepared by U.S. EPA, Washington DC, 20460 SORI-EAS-85-841, November 1985

Blanchard D., Atten P., Dumitran L.M. (2002). Correlation between current density and layer structure for fine particle deposition in a laboratory electrostatic precipitator, *IEEE Transaction on Industry Applications*, Vol. 38, no. 3 May/June, pp. 832-839.

Brocilo C., Chang J.S., Findlay R.D. (2001). Modeling of electrode geometry effects on dust collection efficiency of wire-plate electrostatics precipitators, *Procee. 8th ICESP*, Vol. 1, Southern Comp. Services Inc., Birmingham, Alabama, USA, A4-3 Series, May 14-17, 2001

Caron A. & Dascalescu L. (2004). Numerical modeling of combined corona – electrostatics fields, *J. of Electrostatics*, Vol. 61, pp. .43-55

Chambers M., Grieco G.J., Caine I.C. (2001). Customized rigid discharge electrodes show superior performance in pulp & paper applications, *Procee. 8th ICESP*, Vol. 1, Birmingham, Alabama, USA, May 14–17, 2001

Chung-Liang Ch. & Hsunling B. (1999). An experimental study on the performance of single discharge wire-plate electrostatic precipitator with back corona, *J. Aerosol Sci.*, Vol. 30, No. 3

Grafender A.M. (2010) Pyły atmosferyczne pod mikroskopem, *Energetyka Cieplna i Zawodowa*, 2/2010, pp. 22-25

Hagemann H. & Ahland E. (1973). Abgasentstaubug von mit Steinkohlenstaub gefeuerten Wasserrohr, *Staub-Reinhalt. Luft,* 33 (1973) Nr. 9, pp. 367-372

Hsunling B., Chungsying L., Chung-Liang Ch. (1994). A model to predict the system performance of an electrostatic precipitator for collecting polydispersed particles, *J. of Air and Waste Manage*, ASSOC, Vol. 45, pp. .908-916

IEEE-DEIS-EHD Technical Committee (2003). Recommended international standard for dimensionless parameters used in electrohydrodynamics, *IEEE Trans. Diel. Electr. Insul.* 10-1, pp. 3-6

Jaworek A., Jędrusik M., Świerczok A., Lackowski M., Czech T., Sobczyk A.T. (2011). Biomass co-firing. New challenge for electrostatic precipitators, *Procce. XII International Conference of Electrostatic Precipitation, ICESP XII*, Nuernberg, 10-13 Mai 2011

Jędrusik M. & Świerczok A. (2006). Experimental test of discharge electrode for collecting of fly ash of different physicochemical properties. *Procee. International Conference on Air Pollution Abatement Technologies – future challenges. ICESP X*, Cairns, Queensland, Australia, 25-29 June 2006,

Jędrusik M. (2008). *Elektrofiltry. Rozwinięcie wybranych technik podwyższania skuteczności odpylania*, Oficyna Wydawnicza Politechniki Wrocławskiej, ISBN 978-83-7493-387-2, Wrocław

Jędrusik M. & Świerczok A. (2009). The influence of fly ash physical & chemical properties on electrostatic precipitation process, *Journal of Electrostatics*, 67, pp. 105-109

Jędrusik M. & Świerczok A. (2011). The influence of unburned carbon particles on electrostatic precipitator collection efficiency, Journal of Physics: Conference Series 301 (2011) 012009, doi:10.1088/1742-6596/301/1/012009

Masuda H., Higashitani K., Yoshida H. (2006). *Powder Technology Handbook*, CRC Press Taylor & Francis Group, ISBN: 1-57444-782-3

Mc Kinney P.J., Davidson J.H., Leone D. M. (1992). Current distributions for barbed plate-to-plane coronas, *IEEE Transaction on industry Applications*, vol. 28, No.6 Nov/Dec, pp. 1424-1431

McCain J.D. (2001). Estimeted Operating V-I curves for rigid frame discharge electrodes for use In ESP modeling, *Procee. 8th ICESP*, Vol. 1, Birmingham, Alabama, USA, May 14–17, 2001

Miller J., Schmid H.J., Schmidt E., Schwab A.J. (1996a). Local deposition of particles in a laboratory-scale electrostatic precipitator with barbed discharge electrodes, *Procee. 6th International Conference on Electrostatic Precipitation*, Budapest, Hungary, 18-21 June 1996

Miller J., Schmidt E., Schwab A.J. (1996b). Improved discharge electrode design yields favourable EHD-field with low dust layer erosion in electrostatic precipitators,

Procee. 6-th International Conference on Electrostatic Precipitation, Budapest, Hungary, 18-21 June 1996

Parker K.R. (1997). *Applied Electrostatic Precipitation*, Blackie Academic & Prof., ISBN 07514 0266 4 London

Pauthenier M.M. & Moreau-Hanot M. (1932) La charge des particules spheriques dans un champ ionize, *Journal de Physique et le Radium*, 3, pp. 590-613

Peek F.W. (1929). *Dielectric phenomena in high voltage engineering*, 3rd ed., MacGraw-Hill, New York

White H.J. (1990). *Industrial Electrostatic Precipitation*, (prep.), International Society for Electrostatic Precipitation, Library of Congress Catalog Card No. 62-18240

Permissions

The contributors of this book come from diverse backgrounds, making this book a truly international effort. This book will bring forth new frontiers with its revolutionizing research information and detailed analysis of the nascent developments around the world.

We would like to thank Dr. Mukesh Khare , for lending his expertise to make the book truly unique. He has played a crucial role in the development of this book. Without his invaluable contribution this book wouldn't have been possible. He has made vital efforts to compile up to date information on the varied aspects of this subject to make this book a valuable addition to the collection of many professionals and students.

This book was conceptualized with the vision of imparting up-to-date information and advanced data in this field. To ensure the same, a matchless editorial board was set up. Every individual on the board went through rigorous rounds of assessment to prove their worth. After which they invested a large part of their time researching and compiling the most relevant data for our readers. Conferences and sessions were held from time to time between the editorial board and the contributing authors to present the data in the most comprehensible form. The editorial team has worked tirelessly to provide valuable and valid information to help people across the globe.

Every chapter published in this book has been scrutinized by our experts. Their significance has been extensively debated. The topics covered herein carry significant findings which will fuel the growth of the discipline. They may even be implemented as practical applications or may be referred to as a beginning point for another development. Chapters in this book were first published by InTech; hereby published with permission under the Creative Commons Attribution License or equivalent.

The editorial board has been involved in producing this book since its inception. They have spent rigorous hours researching and exploring the diverse topics which have resulted in the successful publishing of this book. They have passed on their knowledge of decades through this book. To expedite this challenging task, the publisher supported the team at every step. A small team of assistant editors was also appointed to further simplify the editing procedure and attain best results for the readers.

Our editorial team has been hand-picked from every corner of the world. Their multi-ethnicity adds dynamic inputs to the discussions which result in innovative outcomes. These outcomes are then further discussed with the researchers and contributors who give their valuable feedback and opinion regarding the same. The feedback is then collaborated with the researches and they are edited in a comprehensive manner to aid the understanding of the subject.

Apart from the editorial board, the designing team has also invested a significant amount of their time in understanding the subject and creating the most relevant covers. They scrutinized every image to scout for the most suitable representation of the subject and create an appropriate cover for the book.

The publishing team has been involved in this book since its early stages. They were actively engaged in every process, be it collecting the data, connecting with the contributors or procuring relevant information. The team has been an ardent support to the editorial, designing and production team. Their endless efforts to recruit the best for this project, has resulted in the accomplishment of this book. They are a veteran in the field of academics and their pool of knowledge is as vast as their experience in printing. Their expertise and guidance has proved useful at every step. Their uncompromising quality standards have made this book an exceptional effort. Their encouragement from time to time has been an inspiration for everyone.

The publisher and the editorial board hope that this book will prove to be a valuable piece of knowledge for researchers, students, practitioners and scholars across the globe.

List of Contributors

Pehoiu Gica and Murărescu Ovidiu
"Valahia" University of Târgovişte, Romania

Sumanth Chinthala and Mukesh Khare
Indian Institute of Technology Delhi, India

Mehmet Yaman
Firat University, Science Faculty, Department of Chemistry, Elazig, Turkey

S. Despiau
LSEET-LEPI / UMR 6017/ Université du Sud Toulon-Var, France

Yasser Antonio Fonseca Rodríguez, Elieza Meneses Ruiz and Gil Capote Mastrapa
CUBAENERGIA, Cuba

Turtós Carbonell and José de Jesús Rivero Oliva
Universidade Federal do Rio de Janeiro, Brazil

Mehran Hoodaji, Mitra Ataabadi and Payam Najafi
Islamic Azad University, Khorasgan Branch (Isfahan), Iran

Gabriele Curci
Dept. Physics, CETEMPS, University of L'Aquila, Italy

Arun Kumar Sharma, Shveta Acharya, Rashmi Sharma and Meenakshi Saxena
Department of Chemistry, S.D. Govt. College, Beawar, Rajasthan, India

Barbara Majkowska-Wojciechowska and Marek L. Kowalski
Department of Immunology, Rheumatology and Allergy, Medical University of Łódź, Poland

Maria Jędrusik and Arkadiusz Świerczok
Wroclaw University of Technology, Poland

Printed in the USA
CPSIA information can be obtained
at www.ICGtesting.com
JSHW011441221024
72173JS00004B/889